THE ELECTRICIAN'S TROUBLESHOOTING POCKET GUIDE

THE ELECTRICIAN'S TROUBLESHOOTING POCKET GUIDE

John E. Traister

McGraw-Hill

New York San Francisco Washington, D.C. Auckland Bogotá
Caracas Lisbon London Madrid Mexico City Milan
Montreal New Delhi San Juan Singapore
Sydney Tokyo Toronto

McGraw-Hill

A Division of The **McGraw-Hill** *Companies*

2 3 4 5 6 7 8 9 0 DOC/DOC 9 0 1 0 9 8 7

ISBN-0-07-065323-2

National Electrical Code® and *NEC*® are registered trademarks of the National Fire Protection Association, Inc., Quincy, MA 02269.

Printed and bound by R.R. Donnelley/Crawfordsville.

This book is printed on acid-free paper.

Contents

PART 1
Electrical Essentials

PART 2
Troubleshooting and Maintenance of Electrical Systems

INTRODUCTION

Electrical troubleshooting plays an important role in the installation and maintenance of electrical systems of all types. Therefore, technicians who are involved in the installation, maintenance, and repair of electrical systems should have a good working knowledge of the various testing instruments available and how to use them.

The use of electrical/electronic testing instruments generally involves three applications:

1. Verifying proper operation of instruments and associated equipment.

2. Calibrating electronic instruments and associated equipment.

3. Troubleshooting electrical/electronic circuits and equipment.

For these applications, specific test equipment is selected to analyze circuits and to determine specific characteristics of discrete components.

The test equipment that an electrical technician chooses for a specific task depends on the type of measurement and level of accuracy required. Additional factors that may influence selection include the following:

- Whether the test equipment is portable.
- The amount of information that the test equipment provides.
- The likelihood that the test equipment may damage the circuit or component being tested (some test equipment can generate enough voltage or current to damage an instrument or electrical circuit).

This guide focuses on troubleshooting techniques that electrical technicians will be required to use on the job. The intent is to familiarize you with the use and operation of such equipment and to provide practical applications involving such equipment. This guide further helps in selecting the appropriate test equipment and then effectively using the equipment to perform a required testing or troubleshooting task.

The latter part of this book contains dozens of troubleshooting charts that will help electrical technicians deal with common malfunctions of electrical equipment and components. Consequently, this book should prove useful to anyone who is involved in the electrical industry in any capacity.

John E. Traister
Bentonville, Virginia

Part 1

ELECTRICAL ESSENTIALS

CHAPTER 1

Troubleshooting Electrical Systems

A great deal of work performed by electrical technicians involves the repair and maintenance of electrical systems. To properly maintain such systems, the workers must have a good knowledge of what is commonly known as troubleshooting — the ability to determine the cause of any electrical problem and then to correct it in the shortest possible time with the least amount of expense or inconvenience to the building or system owners.

Troubleshooting covers a wide range of problems from such small jobs as finding a short circuit or ground fault in a home appliance to tracing out defects in a complex motor-control circuit. In any case, troubleshooting usually requires only a thorough knowledge of electrical theory and testing equipment, and then using a systematic and methodical approach to the problem; that is, testing one part of the circuit or system after another until the trouble is located.

Those workers involved in the maintenance of electrical systems — from residential to industrial — should keep in mind that every electrical problem can be solved, regardless of the nature. The material covered in this chapter is designed to lay a foundation for electrical technicians to help them better solve the more common electrical problems in a safe and logical manner.

Basic Electrical Problems

In general, there are only three basic electrical faults:

- A short circuit (ground fault)
- An open circuit
- A change in electrical value

A short circuit is probably one of the most common causes of electrical problems. Such a condition can be caused by a number of reasons, but in most cases the cause is faulty insulation on conductors which allows conductors to touch (Figure 1-1) and short out, or else the fault occurs between one ungrounded conductor and a grounded object as illustrated in Figure 1-2.

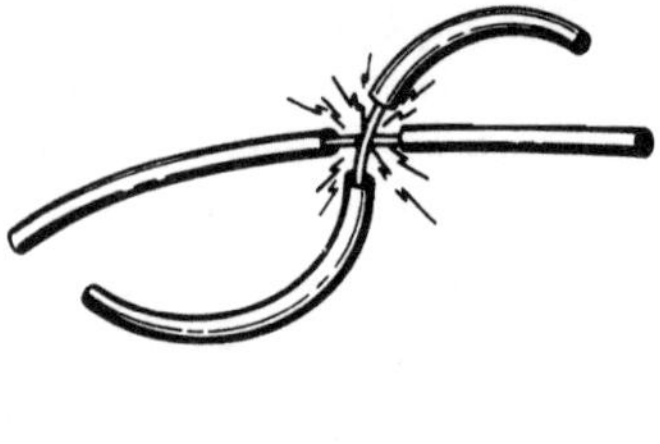

Figure 1-1: Damaged insulation on conductors can cause a short circuit.

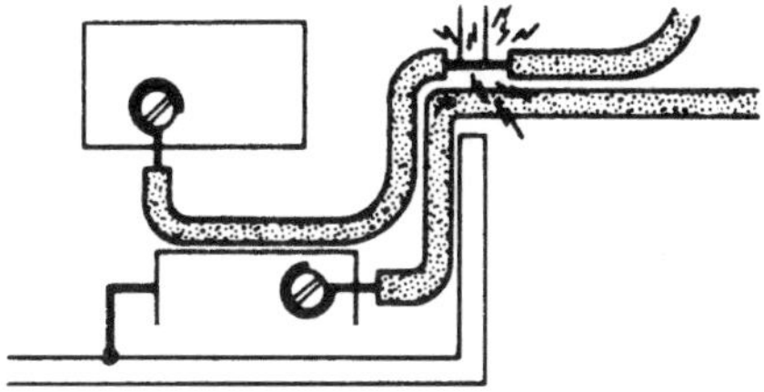

Figure 1-2: A faulty ungrounded conductor touching a grounded frame can cause a short circuit.

An open circuit is an incomplete current path and is usually caused by either a blown or tripped overcurrent-protective device, a loose connection, broken conductor, or a faulty switch or control.

A change in electrical value covers such items as low voltage, electrical surges, a change in resistance, and similar items.

ELECTRICAL ESSENTIALS

Several equations must be used in the electrical technician's daily routine. Most workers are quite familiar with Ohm's law, but many of the basic equations are not needed or used as frequently. Consequently, practical application of these equations soon become hazy until the time comes when they are needed. Then, the technician must hunt an electrical theory book to find out the correct approach. For this and other reasons, a review of electrical theory is covered in this chapter, making this pocket guide even more complete.

Electricity

Electricity is basically the flow of electrons — tiny atomic particles. These particles are found in all atoms. Atoms of some metals such as copper and aluminum have electrons that are easily pushed and guided into a stream. When a coil of metal wire is turned near a magnet, or vice versa, electricity will flow in the wire. This principle is made use of in generating plants; water or steam is used to turn turbines which rotate electromagnets that are surrounded by huge coils of wire. The push transmitted to the electrons by the turbine/magnet setup is measured in units called *volts*. The quantity of the flow of electricity is called *current* and it is measured in *amperes* or *amps*.

Multiply volts by amps and you get *volt-amperes* or *watts* — the power or amount of work that electricity can do. Electrical appliances and motors have certain wattage requirements depending on the task they are expected to perform. For convenience, we can use *kilowatts*, (1 kW equals 1000 W) when speaking of power production or power needs. A power plant produces kilowatts which are sold to users by the *kilowatt-hour.* For example, a 100-W lamp left on for ten hours uses 1 kWh of electricity.

Resistance, or the opposition to the flow of electricity, is another term that will be covered in this chapter. In general, a conductor enhances the flow of electricity while resistance impedes or stops the flow of electricity. Therefore, insulators are constructed of materials offering a high resistance to the flow of electricity.

MEASURING ELECTRICITY

The three basic terms used to measure electricity are:

- Electromotive force — measured in volts
- Current — measured in amperes
- Resistance — measured in ohms

In general, *electromotive force* or *voltage* is the force that causes electrons to flow. The unit of measurement of voltage is volts. In equations, voltage is represented by the letter E which stands for electromotive force. Voltage is measured with a *voltmeter.*

Current is the rate at which electrons flow in a circuit. A current of 1 A is said to flow when 1 C of charge passes a point in 1 second. One coulomb is equal to the charge of 6.28×10^{18} electrons. Current is measured in units called amperes or amps. In equations, current is represented by the letter *I* which stands for intensity of current.

In many cases, the ampere is too large a unit for measuring current. Therefore, the *milliampere* (mA) or the *microampere* (μA) is used. A milliampere equals one-thousandth of an ampere, while the microampere equals one-millionth of an ampere. The device used to measure current is called an *ammeter*.

Resistance is the opposition to the flow of current in a circuit. All circuits have some resistance and the amount of resistance is measured in ohms. In equations, resistance is represented by the letter *R* or the Greek letter omega (Ω). A conductor has one ohm of resistance when an applied potential of one volt produces a current of one ampere.

Resistance, although an electrical property, is determined by the physical structure of a material. The resistance of a material is governed by many of the same factors that control current flow.

Factors that Affect Resistance

The magnitude of resistance is determined in part by the number of free electrons available with the material. Since a decrease in the number of free electrons will decrease the current flow, it can be said that the opposition to current flow (resistance) is greater in a material with fewer free electrons. Consequently, the resistance of a material is determined by the number of free electrons available in a material.

Depending upon their atomic structure, different materials will have different quantities of free electrons. Therefore, the various conductors used in electrical applications have different values of resistance.

Effect of Cross-Sectional Area

Cross-sectional area greatly affects the magnitude of resistance. If the cross-sectional area of a conductor is increased, a greater quantity of electrons are available for movement through the conductor. Therefore, a larger current will flow for a given amount of applied voltage. An increase in current indicates that when the cross-sectional area of a conductor is increased, the resistance must have decreased. If the cross-sectional area of a conductor is decreased, the number of available electrons decreases and, for a given applied voltage, the current through the conductor decreases. A decrease in current flow indicates that when the cross-sectional area of a conductor is decreased, the resistance must have increased.

The diameter of many electrical conductors is often only a fraction of an inch. Therefore, the diameter of conductors is usually expressed in mils (thousandths of an inch). It is also standard practice to assign the unit circular mil to the cross-sectional area of a conductor

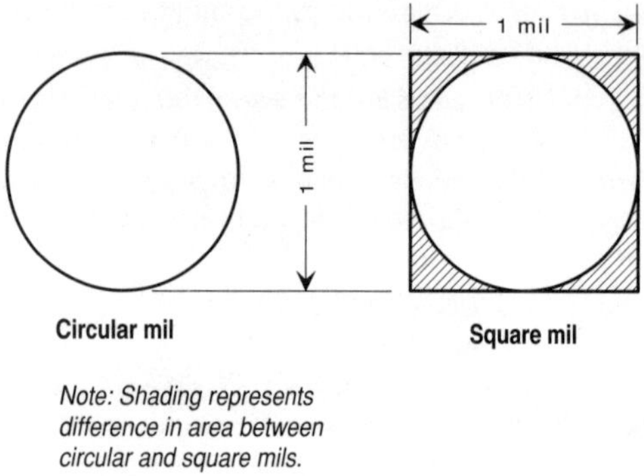

Figure 1-3: Comparison between a square mil and a circular mil.

(See Figure 1-3). The circular mil is found by squaring the diameter, when the diameter is expressed in mils. Consequently, if the diameter of a conductor is, say, 35 mils (0.035 in), the circular mil area is equal to:

$$(35)^2 = 1225 \text{ circular mils}$$

Effect of Conductor Length

The length of a conductor is also a factor which determines the resistance of a conductor. If the length of a conductor is increased, the amount of energy given up increases. As free electrons move from atom to atom, some energy is given off as heat. The longer a conductor is, the more energy is lost to heat. The additional energy loss is subtracted from the energy being transferred through the conductor, resulting in a decrease in current flow for a given applied voltage. A decrease in current flow indicates an increase in resistance, since voltage was held constant. Therefore, if the length of a conductor is increased, the resistance of that conductor increases.

Effect of Temperature

Temperature changes affect the resistance of materials in different ways. In some materials, an increase in temperature causes an increase in resistance, whereas in others, an increase in temperature causes a decrease in resistance. The amount of change of resistance per unit change in temperature is known as the temperature coefficient. When the resistance of a material increases with an increase in temperature, the material is said to have a positive temperature coefficient. When the resistance of a material decreases with an increase in temperature, the material is said to have a negative temperature coefficient. Most conductors used in electrical applications have a positive temperature coefficient. However, carbon — a frequently used material — is a substance having a negative temperature coefficient.

Some materials, such as the alloys constantan and manganin, are considered to have a zero temperature coefficient because their resistance remains relatively constant for changes in temperature.

Conductance

Conductance is the ability of a material to pass electrons, and is the opposite of resistance. The factors that affect the magnitude of resistance are exactly the same for conductance, but they affect conductance in the opposite manner. Therefore, conductance is directly proportional to area, and inversely proportional to the length of the material. The unit of conductance is the MHO, which is ohm spelled backward. Another term for MHO is *siemans*.

DIRECT CURRENT

The common flashlight is an example of a basic electric circuit. It contains a source of electrical energy (the dry cells in the flashlight), a load (the bulb) which changes the electrical energy into light energy, and a switch to control the energy delivered to the load.

The technician's main aid in studying circuits is the schematic diagram. In general, a schematic diagram is a "picture" of the circuit that uses symbols to represent the various circuit components and lines to connect these components. Basic symbols used in schematic diagrams are shown in Figure 1-4. Refer to this symbol list frequently as you review the various schematic diagrams in the remaining pages of this chapter.

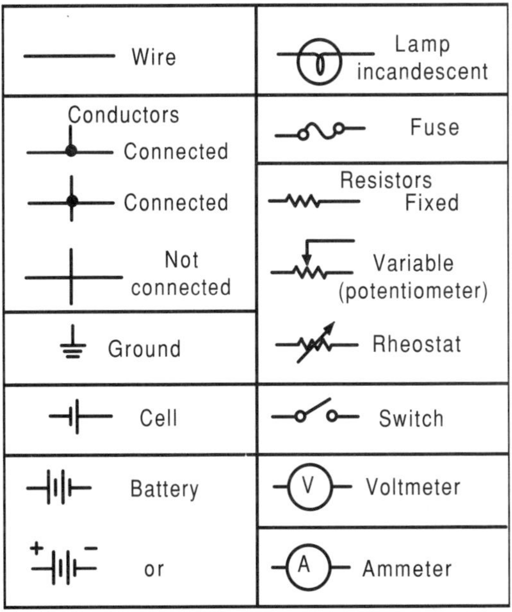

Figure 1-4: Basic symbols used in schematic diagrams.

Figure 1-5 shows a pictorial representation of a common flashlight (A), along with the appropriate symbols, at the approximate locations, for the bulb, switch, and battery. A line is used to connect these components. In actual practice, drafters would not draw the flashlight since we are concerned only with the electrical circuit. Consequently, schematic diagrams are used as shown in (B) and (C) below the pictorial drawing. These are actually simplified representations of the drawing in (A).

Refer again to Figure 1-5 and note the schematic in (B). This drawing shows the flashlight in the OFF or deenergized state. The switch (S1) is open. There is no complete path for current (I) through the circuit, and the bulb (DSI) does not light. In Figure 1-5C, however, switch S1 is closed (moved to the ON position) and current flows in the direction of the arrows from the negative terminal of the battery, through the switch (S1), through the lamp (DS1), and back to the positive terminal of the battery. With the switch closed, the path for current is complete and current will continue to flow until the switch (S1) is moved to the open (OFF) position, or the battery is completely discharged.

Ohm's Law

In the early part of the nineteenth century, George Ohm proved by experimentation that a precise relation-

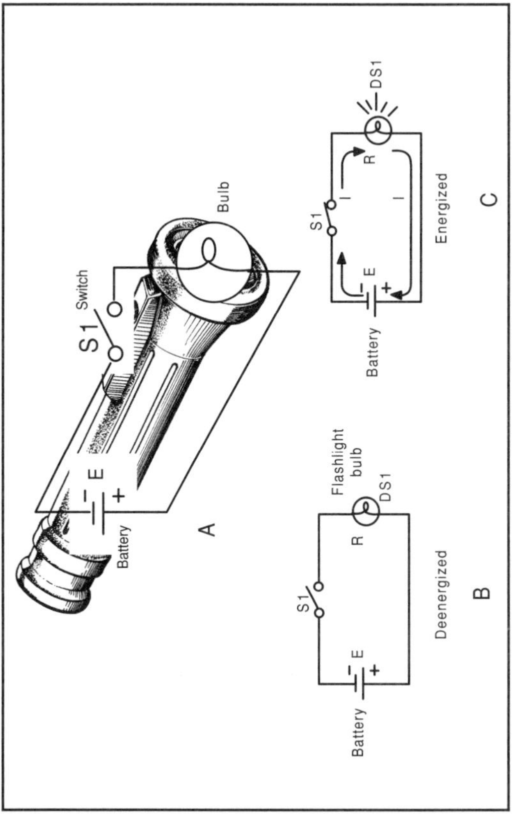

Figure 1-5: Basic electric circuit in a common flashlight.

ship exists between current, voltage, and resistance. This relationship is called *Ohm's law* and is stated as follows:

> *The current in a circuit is directly proportional to the applied voltage and inversely proportional to the circuit resistance.*

Ohm's law may be expressed as an equation:

$$I = \frac{E}{R}$$

where: I = current in amperes

E = voltage in volts

R = resistance in ohms

As stated in Ohm's law, current is inversely proportional to resistance. This means, as the resistance in a circuit increases, the current decreases proportionately.

Refer again to Figure 1-5C — the schematic of the flashlight in the ON state. If the battery supplies a voltage of 1.5 V and the lamp (DS1) has a resistance of 5 Ω, then the current in the circuit can be determined by substituting values in the equation previously described:

$$I = \frac{E}{R} = \frac{1.5\ V}{5\ \Omega} = .3\ A$$

If the flashlight contained two 1.5-V batteries, the voltage would double and 3 V would be applied to the circuit. Using this voltage in the equation:

$$I = \frac{3.0\ V}{5\ \Omega} = .6\ A$$

In comparing these two sample equations, you can see that both the voltage and current has doubled in the latter example. This demonstrates that the current is directly proportional to the applied voltage.

If the value of resistance of the lamp is double, the equation will be:

$$I = \frac{E}{R} = \frac{3.0\ V}{10\ \Omega} = .3\ A$$

The current has been reduced to one-half of the value of the previous equation, or .3 A. This demonstrates that the current is inversely proportinal to the resistance. Doubling the value of the resistance of the load reduces circuit current value to one-half of its former value.

Application of Ohm's Law

In using Ohm's law, if any two of the variables are known, the unknown can be found. For example, if current (I) and voltage (E) are known, resistance (R) can be determined as follows:

Step 1. Use the basic equation:

$$I = \frac{E}{R}$$

Step 2. Remove the divisor by multiplying both sides by R:

$$R \times I = \frac{E}{R} \times \frac{R}{I}$$

Step 3. Note result of Step 2: $R \times I = E$

Step 4. To get R alone, divide both sides by I:

$$\frac{RI}{I} = \frac{E}{I}$$

Step 5. The basic equation, transposed for R, is:

$$R = \frac{E}{I}$$

Now let's put this equation to practical use. Refer to Figure 1-6 and note that the voltage (E) is 10 V, and the current (I) equals 1 A. Solve for R, using the equation just explained.

Step 1. Insert the known values in the equation:

$$R = \frac{10\ V}{1\ A} = 10\ \Omega$$

The basic Ohm's law equation can also be used to solve for voltage (E).

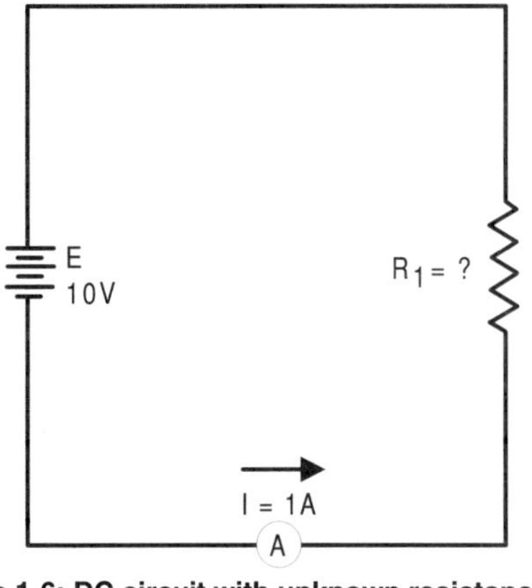

Figure 1-6: DC circuit with unknown resistance.

Step 2. Use the basic equation:

$$I = \frac{E}{R}$$

Step 3. Multiply both sides by R:

$$I \times R = \frac{E}{R} \times \frac{R}{1}$$

Step 4. Note the results of Step 2.

$$E = I \times R$$

Now let's use this equation to find the voltage in the circuit shown in Figure 1-7, where the amperage equals .5 A, and the resistance equals 45 Ω.

$$E = I \times R$$

$$E = .5\,A \times 45\,\Omega$$

$$E = 22.5\ V$$

The Ohm's law equation and its various forms may be readily obtained with the aid of the circle in Figure 1-8. Note that the circle is divided into three parts with each part containing *E*, *I*, and *R*, respectively. The letter *E* is above the horizontal line, while *I* and *R* are below

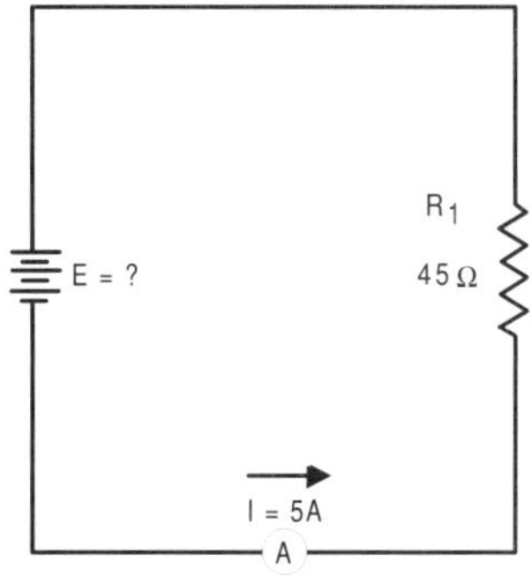

Figure 1-7: DC circuit with unknown voltage.

the horizontal line. To determine the unknown quantity, first cover that quantity with a finger. The position of the uncovered letters in the circle will indicate the mathematical operation to be performed. For example, to find current (*I*), cover the letter *I* with a finger. The uncovered letters indicate that E is to be divided by R, or $I = \frac{E}{R}$. To find the equation for *E*, cover *E* with your finger. The result indicates that I is to be multiplied by *R*, or $E = IR$. To find the equation for *R*, cover *R*. The result indicates that *E* is to be divided by *I*, or $R = \frac{E}{I}$.

Although some shortcut methods save time when used on the job or in practical applications, the technician should have a basic knowledge of how these shortcut

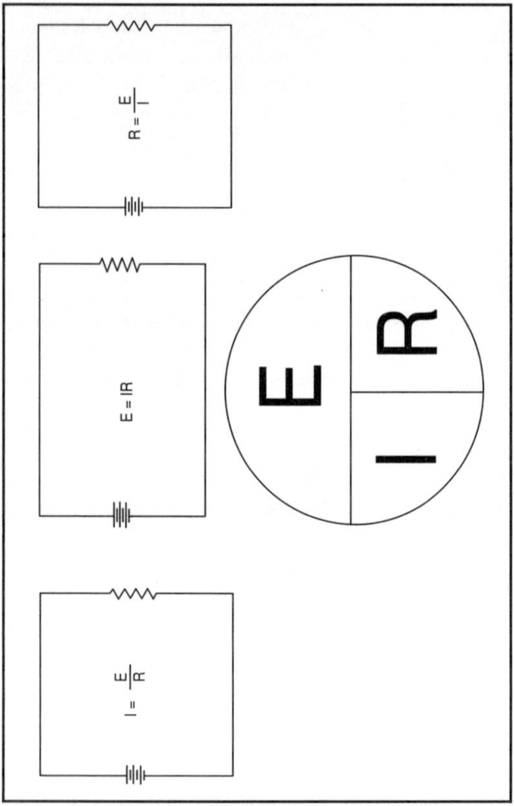

Figure 1-8: Ohm's law in diagram form.

methods are obtained. The diagram in Figure 1-8 should be used to supplement your knowledge of the algebraic method, not as a substitution for the algebraic method. Algebra is a basic tool in the solution of electrical problems.

Power

Power, whether electrical or mechanical, pertains to the rate at which work is being done. Work is done whenever a force causes motion. When a mechanical force is used to lift or move a weight, work is done. When voltage causes electrons to move, work is done. The instantaneous rate at which this work is done is called the electric power rate, and is measured in watts or volt-amperes. Power in watts is equal to the voltage across a circuit multiplied by current through the circuit. This represents the rate at any given instant at which work is being done. The symbol P indicates electrical power, and the basic power equation is:

$$P = E \times I$$

where: E = voltage and I = current.

The amount of power changes when either voltage or current, or both voltage and current, are caused to change. The power equation also has variations similar to those discussed previously for determining either voltage, current, or resistance in a circuit.

Electrical components are often given a power rating. The power rating, in watts, indicates the rate at which the device converts electrical energy into another form of energy, such as light, heat, or motion.

As with other electrical quantities, prefixes may be attached to the word *watt* when expressing very large or very small amounts of power. Some common examples of these are the kilowatt (1000 W), the megawatt (1,000,000 W), and the milliwatt ($^1/_{1000}$ of a watt). The National Electrical Code® is now referring to watts as "volt-amperes" in most NEC Articles and Sections, so the technician should be aware that power may be expressed in either watts or volt-amperes (VA).

Four of the most important electrical quantities have been discussed thus far:

- Voltage (E)
- Current (I)
- Resistance (R)
- Power (P)

Anyone involved in the electrical industry in any capacity must understand the relationships that exist among these quantities because they are used throughout this book and will be used throughout the technician's career. Figure 1-9 is a summary of 12 basic equations

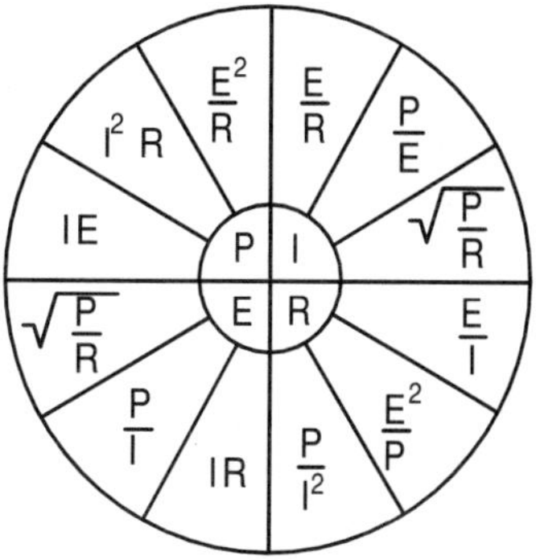

Figure 1-9: Summary of basic electrical equations.

that you should know. The four quantities E, I, R, and P are at the center of the circle. Adjacent to each quantity are three segments. Note that in each segment, the basic quantity is expressed in terms of two other basic quantities, and no two segments are alike.

SERIES DC CIRCUITS

A series circuit is defined as a circuit that contains only one path for current flow. To compare the basic flashlight circuit (discussed previously) with a more complex series circuit, refer to Figure 1-10. Remember that our basic flashlight circuit had only one lamp, while the more complex series circuit in Figure 1-10 contains three lamps connected in series. The current in this circuit must flow through each lamp to complete the electrical path in the circuit. Each additional lamp offers added resistance. Consequently, in a series circuit, the total circuit resistance (R_T) is equal to the sum of the individual resistances.

$$R_T = R_1 + R_2 + R_3 + \ldots + R_n$$

The series circuit in Figure 1-10 consists of three resistors; one rated at 10 Ω, one at 15 Ω, and one at 30 Ω. A voltage source provides 120 V. What is the total resistance?

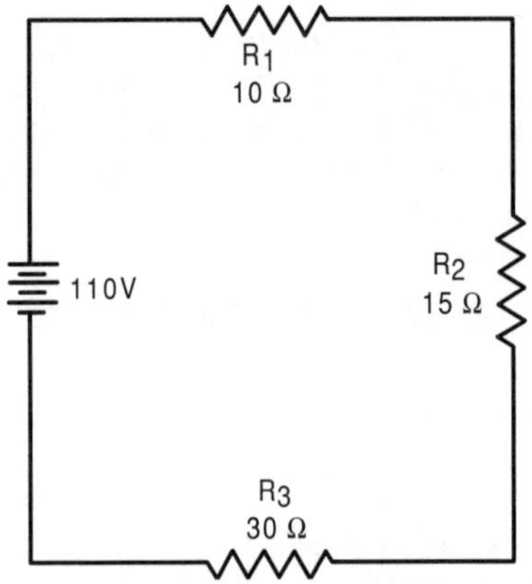

Figure 1-10: Solving for total resistance in a series circuit.

Step 1. Use the basic equation for finding resistance in a series circuit.

Step 2. Substitute known values in the equation.

$$R_T = 10\ \Omega + 15\ \Omega + 30\ \Omega$$

$$R_T = 55\ \Omega$$

In some applications, the total resistance is known and the value of one of the circuit resistors has to be determined. The former equation for finding resistance in a series circuit can be transposed to solve for the value of the unknown resistance.

For example, the series circuit in Figure 1-11 on the next page has a total resistance of 40 Ω. Two of the resistors are rated at 10 Ω each, while the rating of resistor R_3 is unknown. Here's how to find the value of the unknown resistor.

Step 1. Use the basic equation.

Step 2. Subtract $R_1 + R_2$ from both sides of the equation.

Step 3. Continue solving for R_3 as follows:

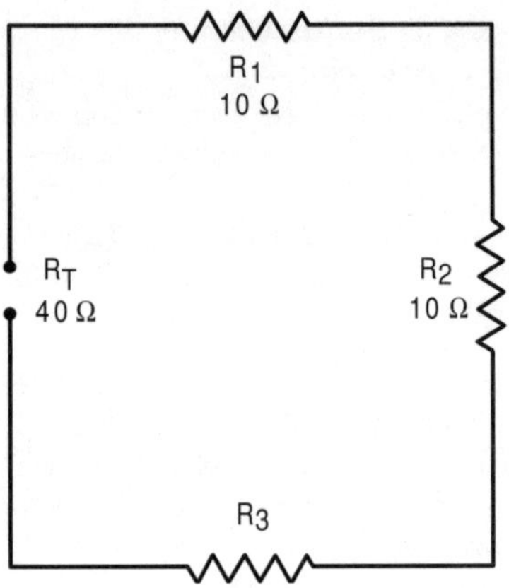

Figure 1-11: Calculating the value of one unknown resistance in a series circuit.

$$R_T - R_1 - R_2 = R_3$$

$$R_3 = R_T - R_1 - R_2$$

$$R_3 = 40\ \Omega - 10\ \Omega - 10\ \Omega$$

$$R_3 = 40\ \Omega - 20\ \Omega$$

$$R_3 = 20\ \Omega$$

Current in a Series Circuit

Since there is only one path for current in a series circuit, the same current must flow through each component of the circuit. To determine the current in a series circuit, only the current through one of the components need be known. Ohm's law may be used to calculate the current in a series circuit if the voltage and resistance quantities are known.

The current flow through each component of a series circuit can be verified by inserting meters into the circuit at various points, as shown in Figure 1-12. Upon examining these meters, each meter would be found to indicate the same value of current.

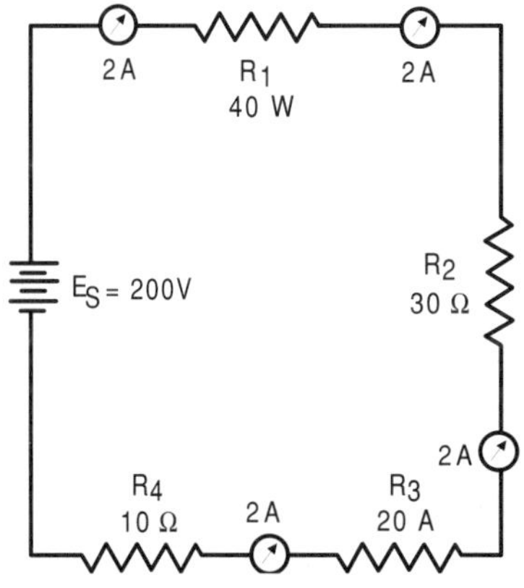

Figure 1-12: Current in a series circuit.

Voltage in a Series Circuit

The voltage drop across the resistor in a circuit consisting of a single resistor and a voltage source is the total voltage across the circuit and is equal to the applied voltage. The total voltage across a series circuit that consists of more than one resistor is also equal to the applied voltage, but consists of the sum of the individual resistor voltage drops.

In any series circuit, the sum of the resistor voltage drops must equal the source voltage. This statement can be proven by examining the series circuit in Figure 1-13. In this circuit, a source voltage (E_T) of 20 V is dropped

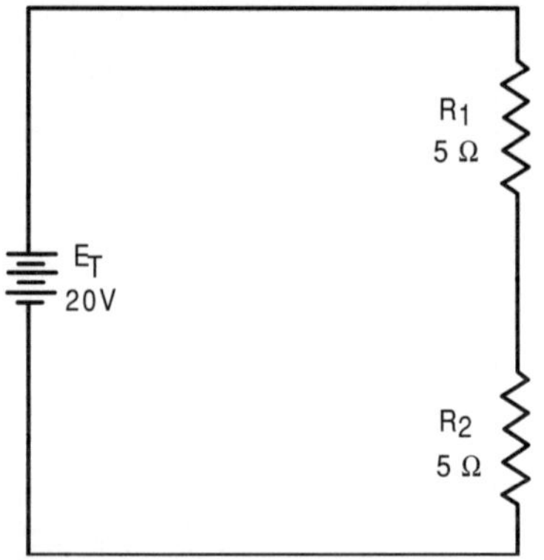

Figure 1-13: Calculating individual voltage drop in a series circuit.

across a series circuit consisting of two 5-Ω resistors. The total resistance of the circuit (R_T) is equal to the sum of the two individual resistances, or 10 Ω. Using Ohm's law, the circuit current may be calculated as follows:

$$I_T = \frac{E_T}{R_T}$$

$$I_T = \frac{20\ V}{10\ \Omega}$$

$$I_T = 2\ A$$

Since the value of the resistors is known to be 5 Ω each, and the current through the resistors is known to be 2 A, the voltage drops across the resistors can be calculated as follows:

$$E_1 = I_1 \times R_1$$

$$E_1 = 2\ A \times 5\ \Omega$$

$$E_1 = 10\ V$$

Since R_2 is the same ohmic value as R_1, and carries the same current, the voltage drop across R_2 also equals 10 V. Then, adding the voltage drops for R_1 and R_2, we obtain (10 + 10 =) 20 V, which is equal to the applied voltage in the circuit. From the previous explanation, we see that the total voltage in a dc series circuit may be obtained by using the following equation:

$$E_T = E_1 + E_2 + E_3 + \ldots + E_n$$

To use the above equation on a practical application, let's assume that a circuit has three resistors connected in series of 20, 30, and 50 Ω, respectively. An ammeter shows that 2 A of current flows through the circuit. What is the total applied voltage?

Step 1. Draw a circuit diagram as shown in Figure 1-14.

Step 2. Solve for the voltage drop across resistor R_1.

$$E_1 = R_1 \times I$$

$$E_1 = 20\ \Omega \times 2\ \text{A} = 40\ \text{V}$$

Step 3. Solve for the voltage drop across resistor R_2.

$$E_2 = R_2 \times I$$

$$E_2 = 30\ \Omega \times 2\ \text{A} = 60\ \text{V}$$

Step 4. Solve for the voltage drop across resistor R_3.

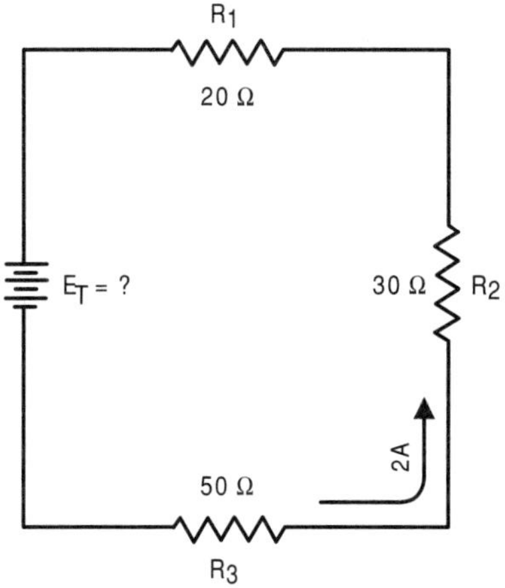

Figure 1-14: Solving for applied voltage in a series circuit.

$$E_3 = R_3 \times I$$

$$E_3 = 50\ \Omega \times 2\ A = 100\ V$$

Step 5. Add the three voltage drops.

$$E_T = 40\ V + 60\ V + 100\ V$$

$$E_T = 200\ V$$

Power in a Series Circuit

Each of the resistors in a series circuit consumes power which is dissipated in the form of heat. Since this power must come from the source, the total power must be equal to the power consumed by the circuit resistances. In a series circuit, the total power is equal to the sum of the power dissipated by the individual resistors. The equation to find the total power in a series circuit follows:

$$P_T = P_1 + P_2 + P_3 + \ldots + P_n$$

Let's determine the total power in watts for the series circuit in Figure 1-15. Note that this circuit has an applied potential of 120 V and three resistors are connected in series, each rated at 5 Ω, 10 Ω, and 15 Ω, respectively.

Step 1. Find the total resistance in the circuit.

$$R_T = R_1 + R_2 + R_3$$

$$R_T = 5\ \Omega + 10\ \Omega + 15\ \Omega$$

$$R_T = 30\ \Omega$$

Step 2. Determine the circuit current.

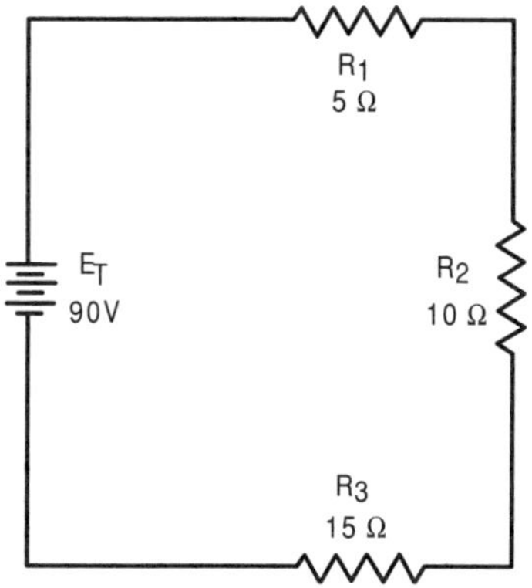

Figure 1-15: Solving for total power in a series circuit.

$$I = \frac{E_T}{R_T}$$

$$I = \frac{120\,V}{30\,\Omega}$$

$$I = 4\text{ A}$$

Step 3. Use the power equation to calculate the power for each resistor.

$$P_{1(2)(3)} = I^2 \times R_{1(2)(3)}$$

Step 4. Calculate the power for resistor R_1.

$$P_1 = (4\text{ A})2 \times 5\ \Omega$$

$$P_1 = 80\text{ W}$$

Step 5. Calculate the power for resistors R_2 and R_3 using the same steps as given in Step 4.

$$P_2 = 160\text{ W}$$

$$P_3 = 240\text{ W}$$

Step 6. Obtain the total power by adding all power values.

$$P_T = 80\text{ W} + 160\text{ W} + 240\text{ W}$$

$$P_T = 480\text{ W}$$

When the total source voltage and the total source amperage are known, the total wattage may be found by multiplying the volts times the amps (volt-amperes). Let's check the previous example with this method.

$$P_{Source} = E_{Source} \times I_{Source}$$

$$P_{Source} = 120\text{ V} \times 4\text{ A}$$

$$P_{Source} = 480\text{ VA (W)}$$

KIRCHHOFF'S VOLTAGE LAW

In 1847, G. R. Kirchhoff extended the use of Ohm's law by developing a simple concept concerning the voltages contained in a series circuit loop. Kirchhoff's law states:

> *The algebraic sum of the voltage drops in any closed path in a circuit and the electromotive forces in that path are equal to zero.*

Kirchhoff's voltage law can be written as an equation as follows:

$$E_a + E_b + E_c + \ldots + E_n = 0$$

where E_a, E_b, etc., are the voltage drops around any closed circuit loop.

To set up the equation for an actual circuit, the following procedure is used:

Step 1. Assume a direction of current through the circuit.

Step 2. Using the assumed direction of current, assign polarities to all resistors through which the current flows.

Step 3. Place the correct polarities on any sources included in the circuit.

Step 4. Starting at any point in the circuit, trace around the circuit, writing down the amount and polarity of the voltage across each component in succession. The polarity used is the sign after the assumed current has passed through the component. Stop when the point at which the trace was started is reached.

Step 5. Place these voltages, with their polarities, into the equation and solve for the desired quantity.

To place the above procedures in use, assume that three resistors are connected in series with a 50-volt source. What is the voltage across the third resistor if the voltage drops across the first two resistors are 25 V and 15 V, respectively?

Step 1. Draw a diagram such as the one shown in Figure 1-16.

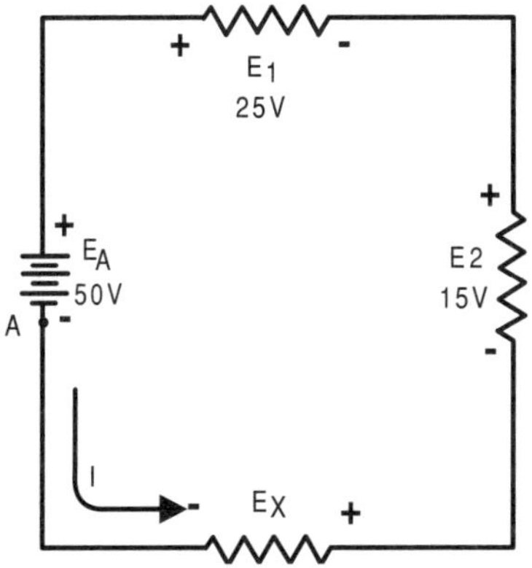

Figure 1-16: Determining unknown voltage in a series circuit.

Step 2. Draw an arrow indicating the assumed direction of current flow. Again, see Figure 1-16.

Step 3. Using the current direction arrow as made in Step 2, mark the polarity (– or +) at each end of each resistor and also on the terminals of the source. These markings are also shown in Figure 1-16.

Step 4. Starting at point A, trace around the circuit in the direction of current flow, recording the voltage and polarity of each component.

Step 5. Starting at point A and using the components from the circuit, we have:

$$(+E_x) + (+E_2) + (+E_1) + (-E_a) = 0$$

Step 6. Substitute known values in the equation from the circuit.

$$E_X + 15\text{ V} + 25\text{ V} - 50\text{ V} = 0$$

$$E_X - 10\text{ V} = 0$$

$$E_X = 10\text{ V}$$

The unknown voltage (E_X) is found to be 10 V.

Solving for Unknown Current

Using the same procedure as above, problems may be solved in which the current is the unknown quantity.

For example, let's assume that a series circuit has a source voltage of 60 V and contains three resistors of 5 Ω, 10 Ω, and 15 Ω. Find the circuit current.

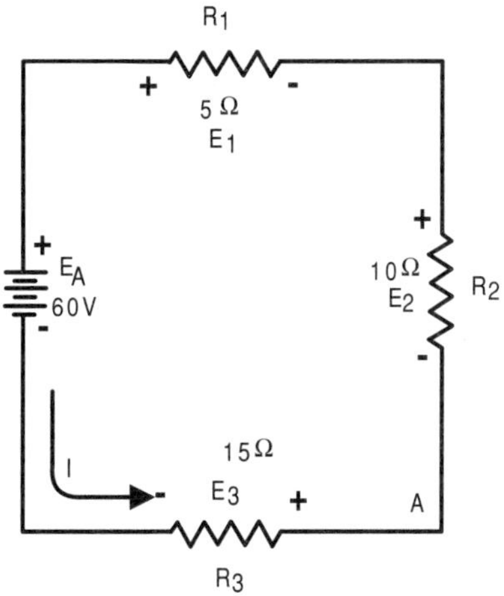

Figure 1-17: Correct direction of assumed current.

Step 1. Draw and label the circuit as shown in Figure 1-17.

Step 2. Start at any point and write out the loop equation.

$$E_2 + E_1 + E_A + E_3 = 0$$

Step 3. Since E = IR, substitute known values in the equation.

$$(I \times R_2) + (I \times R_1) + E_A + (I \times R_3) = 0$$

$$(I \times 10\ \Omega) + (I \times 5\ \Omega) + (-60\ V) + (I \times 15\ \Omega) = 0$$

Step 4. Combine like terms.

$$(I \times 30\ \Omega) + (-60\ V) = 0$$

$$I = \frac{60\ V}{30\ \Omega}$$

$$I = 2\ A$$

Since the current obtained in the preceding calculation is a positive 2 A, the assumed direction of current was correct. However, if the calculation had been a negative value, the assumed direction of current flow would be incorrect. Even if the wrong current direction is assumed, the amount of current in the calculation will be the same. The polarity, however, is negative if the wrong current direction is chosen. In this case, all that is required is to reverse the direction of the assumed current flow. However, should it be necessary to use this negative current value in further calculations on the circuit using Kirchhoff's law, the negative polarity should be retained in the calculations.

PARALLEL DC CIRCUITS

A *parallel circuit* is defined as one having more than one current path connected to a common voltage source. Parallel circuits, therefore, must contain two or more resistances which are not connected in series. An example of a basic parallel circuit is shown in Figure 1-18.

Start at the voltage source (ES) and trace counterclockwise around the circuit. Two complete and separate paths can be identified in which current can flow. One path is traced from the source, through resistance R_1, and back to the source. The other path is from the source, through resistance R_2, and back to the source.

Voltage in a Parallel Circuit

The source voltage in a series circuit divides proportionately across each resistor in the circuit. However, in a parallel circuit, the same voltage is present in each branch. In Figure 1-18, this voltage is equal to the applied voltage (E_S) and can be expressed in the following equation:

$$E_s = E_{R1} = E_{R2}$$

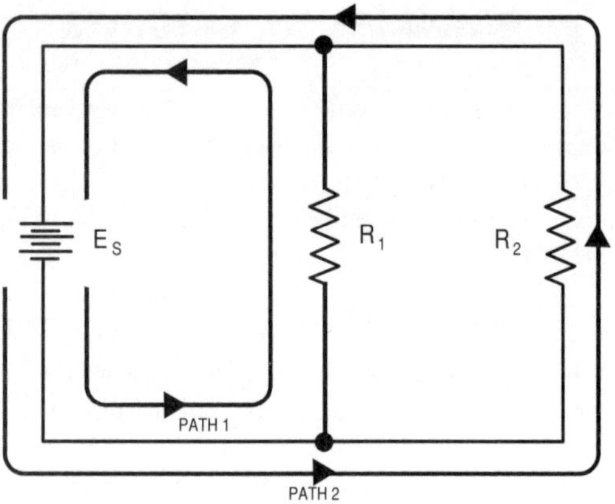

Figure 1-18: A basic parallel circuit.

Voltage measurements taken across the resistors of a parallel circuit are illustrated in Figure 1-19. Each voltmeter indicates the same amount of voltage. Also note that the voltage across each resistor in the circuit is the same as the applied voltage.

For example, assume that the current through a resistor of a parallel circuit is known to be 4.5 mA and the value of the resistor is 30,000 Ω (30 kΩ). What is the source voltage?

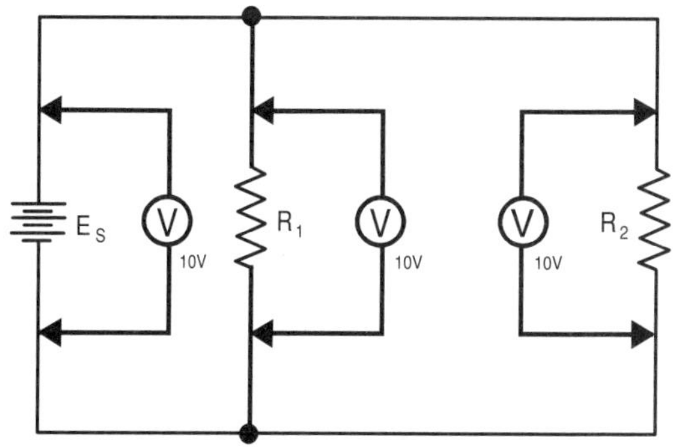

Figure 1-19: Voltage comparison in a parallel circuit.

The circuit in question is shown in Figure 1-20 on the next page and the source voltage may be found by using the basic Ohm's law equation:

$$E = IR$$

Substituting the known values in the equation, we have the following:

$$E_{R2} = .0045 \text{ amp} \times 30{,}000\ \Omega$$

$$E_{R2} = 135 \text{ V}$$

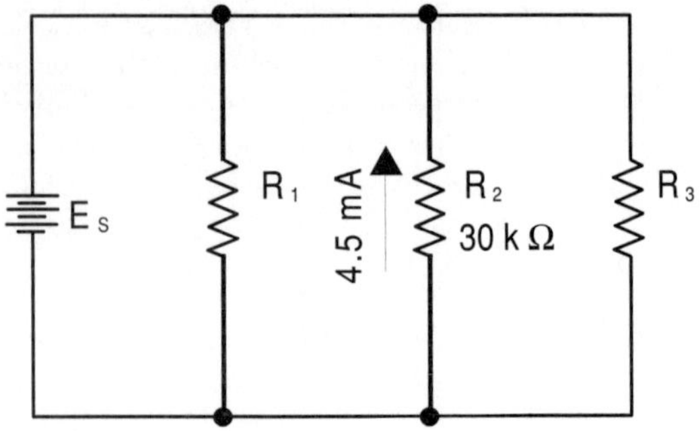

Figure 1-20: Finding source voltage in a parallel circuit.

Since the source voltage is equal to the voltage of a branch, 135 V is the source voltage and is also the voltage applied to each branch of this circuit.

Current in a Parallel Circuit

Ohm's law states that the current in a circuit is inversely proportional to the circuit resistance. This fact is true in both series and parallel circuits.

There is a single path for current in a series circuit. The amount of current is determined by the total resistance of the circuit and the applied voltage. In a parallel circuit, the source current divides among the available paths.

Part (A) of Figure 1-21 shows a basic series circuit. Here, the total current must pass through the single resistor (R_1). Note that the applied voltage equals 50 V and the resistance of R_1 is 10 Ω. The amount of current can be determined by using Ohm's law and is calculated as follows:

$$I = \frac{E}{R}$$

$$I_T = \frac{50\ V}{10\ \Omega}$$

$$I_T = 5\ A$$

Part (B) of Figure 1-21 shows the same resistor (R_1) with a second resistor (R_2) of equal value connected in parallel across the voltage source. When Ohm's law is applied, the current flow through each resistor is found to be the same as the current through the single resistor in part (A).

$$I = \frac{E}{R}$$

$$ES = ER1 = ER2$$

$$I_{R1} = \frac{50\ V}{10\ \Omega}$$

$$I_{R1} = 5\ A$$

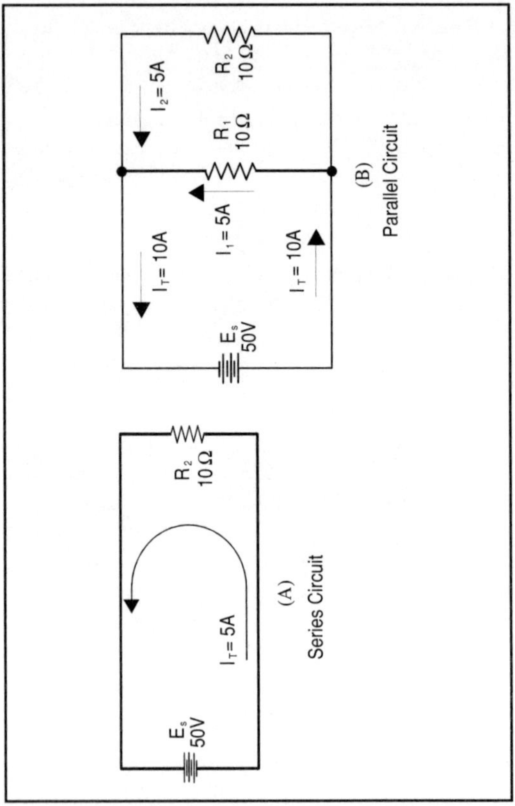

Figure 1-21: Analysis of current in a series and parallel circuit.

$$I_{R2} = \frac{50\ V}{10\ \Omega}$$

$$I_{R2} = 5\ A$$

It is apparent that if there are 5 A of current through each of the two resistors, there must be a total current of 10 A drawn from the source.

The total current of 10 A leaves the negative terminal of the battery and flows to point a. Since point a is a connecting point for the two resistors, it is called a junction. At junction a, the total current divides into two currents of 5 A each. These two currents flow through their respective resistors and rejoin at junction b. The total current then flows from junction b back to the positive terminal of the source. The source supplies a total current of 10 A and each of the two equal resistors carries one-half of the total current.

Each individual current path in the circuit in Figure 1-21(B) is referred to as a *branch*. Each branch carries a current that is a portion of the total current. Two or more branches form a *network*.

From the previous explanation, the characteristics of current in a parallel circuit can be expressed in terms of the following general equation:

$$I_T = I_1 + I_2 + \ldots + I_n$$

Kirchhoff's Current Law

The division of current in a parallel network follows a definite pattern. This pattern is described by *Kirchhoff's current law* which states:

> *The algebraic sum of the currents entering and leaving any junction of conductors is equal to zero.*

This law can be stated mathematically as:

$$I_a + I_b + \ldots + I_n = 0$$

where: I_a, I_b, etc. are the currents entering and leaving the junction. Currents entering the junction are considered to be positive and currents leaving the junction are considered to be negative. When solving a problem using Kirchhoff's current law, the currents must be placed into the equation with the proper polarity signs attached.

Now let's use Kirchhoff's current law to solve for the value of I_3 in Figure 1-22.

The known values are first substituted in Kirchhoff's current law equation.

$$I_1 + I_2 + I_3 + I_4 = 0$$

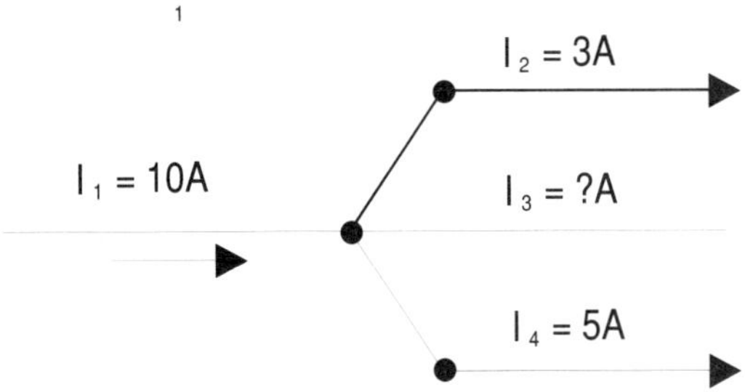

Figure 1-22: Circuit with four current values.

$$10 \text{ amps} + (-3 \text{ A}) + I_3 + (-5 \text{ A}) = 0$$

$$I_3 + 2 \text{ A} = 0$$

$$I_3 = -2 \text{ A}$$

Current I_3 has a value of 2 A, and the negative sign shows it to be a current leaving the junction.

Resistance In Parallel Circuits

Figure 1-23 on the next page shows two resistors connected in parallel across a 5-volt battery. Each has a resistance value of 10 Ω. A complete circuit consisting of two parallel paths is formed and current flows as shown.

Computing the individual currents show 0.5 A flows through each resistor. The total current flowing from the battery to the junction of the resistors, and returning from the resistors to the battery, is equal to 1 A.

The total resistance of the circuit can be calculated by using the values of total voltage (E_T) and total current (I_T).

$$R = \frac{E}{I}$$

$$R_T = \frac{5\ V}{1\ A}$$

$$R_T = 5\ \Omega$$

This computation shows the total resistance to be 5 Ω; one-half the value of either of the two resistors.

The total resistance of a parallel circuit is smaller than any of the individual resistors.

In other words, the total resistance in a parallel circuit is not the sum of the individual resistor values as was the case in a series circuit.

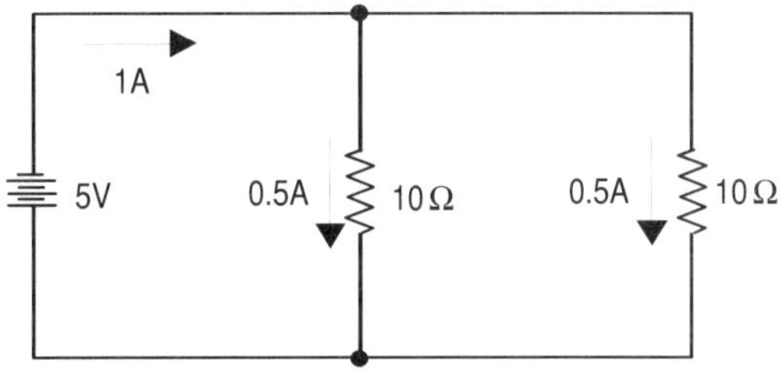

Figure 1-23: Two equal resistors connected in parallel.

There are several methods used to determine the total or equivalent resistance of parallel circuits. The best method for a given circuit depends on the number and value of the resistors. Where all resistors in the circuit are of the same value (Figure 1-23), the following simple equation may be used:

$$R_T = \frac{R}{N}$$

where: R_T = total parallel resistance

R = ohmic value of one resistor

N = number of resistors

The equation is valid for any number of parallel resistors of equal value.

The total resistance of parallel circuits can be found if the individual resistance values are known along with the source voltage. The following equation is the most common:

$$\frac{1}{R_T} = \frac{1}{R_1} + \frac{1}{R_2} + \frac{1}{R_3} + \ldots \frac{1}{R_n}$$

When using the preceding equation to determine the resistance in a parallel circuit, convert the fractions to a common denominator. For example, let's find the total resistance in a parallel circuit with two resistors rated at 3 Ω and 6 Ω, respectively.

$$\frac{1}{R_T} = \frac{1}{3\,\Omega} + \frac{1}{6\,\Omega}$$

$$\frac{1}{R_T} = \frac{2}{6\,\Omega} + \frac{1}{6\,\Omega}$$

$$\frac{1}{R_T} = \frac{3}{6\,\Omega}$$

$$\frac{1}{R_T} = \frac{1}{2\,\Omega}$$

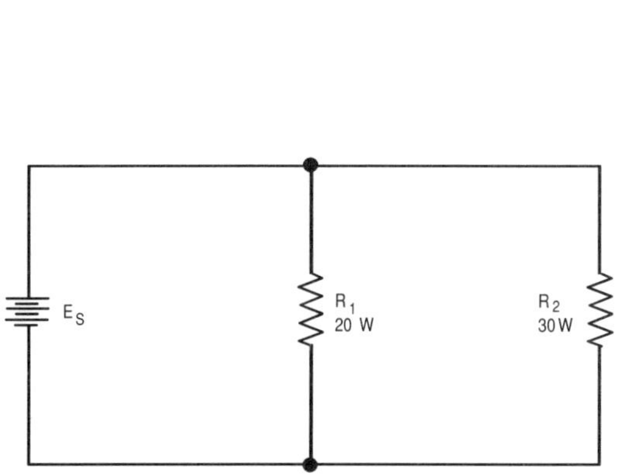

Figure 1-24: Parallel circuit with two unequal resistors.

Since both sides are reciprocals, (divided into one) disregard the reciprocal function.

$$R_T = 2\Omega$$

When only two resistors, each of either the same or different values, are in a parallel circuit, the following equation may be used to find the total resistance in the circuit.

$$R_T = \frac{R_1 \times R_2}{R_1 + R_2}$$

Using the above equation, what is the total resistance in the circuit shown in Figure 1-24?

$$R_T = \frac{20\ \Omega \times 30\ \Omega}{20\ \Omega + 30\ \Omega}$$

$$R_T = \frac{600}{50}\ \Omega$$

$$R_T = 12\ \Omega$$

Equivalent Circuits

In dealing with electrical circuits, it is sometimes necessary to reduce a complex circuit into a simpler form. Any complex circuit consisting of resistances can be redrawn (reduced) to a basic equivalent circuit containing the voltage source and a single resistor representing total resistance. This process is called reduction to an *equivalent circuit.*

Figure 1-25 shows a parallel circuit with three resistors of equal value and the redrawn equivalent circuit. The parallel circuit shown in part (A) shows the original circuit. To create the equivalent circuit, first calculate the total resistance in the circuit.

$$\mathrm{R_T} = \frac{R}{N}$$

$$\mathrm{R_T} = \frac{45\ \Omega}{3}$$

$$\mathrm{R_T} = 15\ \Omega$$

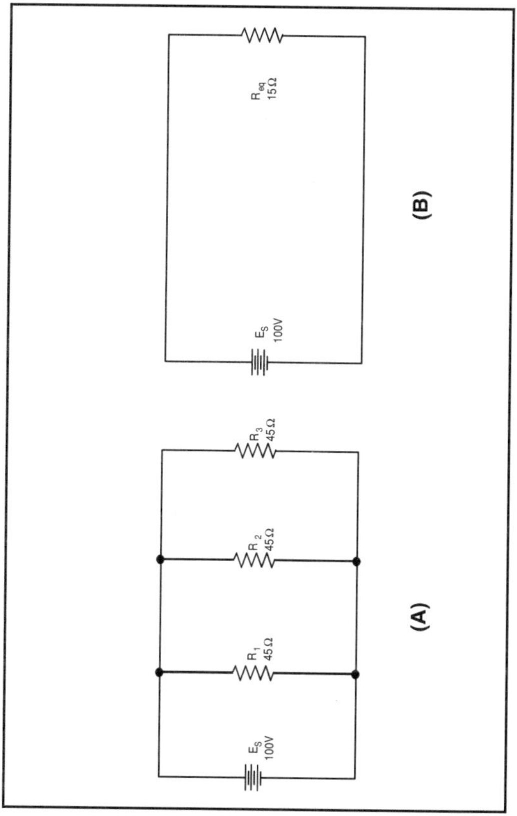

Figure 1-25: Parallel circuit (A) with equivalent circuit (B).

Once the equivalent resistance is known, a new circuit is drawn consisting of a single resistor (to represent the equivalent resistance) and the voltage source, as shown in part (B).

SERIES-PARALLEL DC CIRCUITS

In the preceding sections, series and parallel dc circuits have been considered separately. Electricians often encounter circuits consisting of both series and parallel elements. A circuit of this type is referred to as a *combination circuit.* Solving for the quantities and elements in a combination circuit is simply a matter of applying the laws and rules discussed up to this point.

The basic technique used for solving dc combination-circuit problems is the use of equivalent circuits. To simplify a complex circuit to a simple circuit containing only one load, equivalent circuits are substituted (on paper) for the complex circuit they represent — the technique briefly discussed in the preceding section of this chapter.

To demonstrate the method used to solve combination circuit problems, refer to the circuit in Figure 1-26(A). Examination of this circuit shows that the only quantity that can be computed with the given information is the equivalent resistance of R_2 and R_3. Since only two resistors are contained in this part of the circuit, and since these resistors are connected in parallel, the prod-

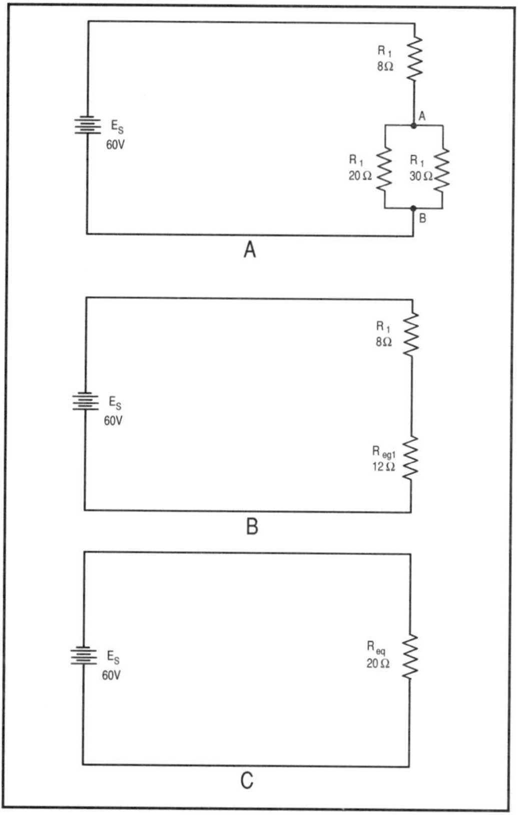

Figure 1-26: Steps in reducing a combination circuit to its simplest form.

uct over the sum equation may be used to obtain the total resistance for this portion of the circuit.

$$R_T = \frac{R_2 \times R_3}{R_2 + R_3}$$

$$R_T = \frac{20\ \Omega \times 30\ \Omega}{20\ \Omega + 30\ \Omega}$$

$$R_T = \frac{600}{50}\ \Omega$$

$$R_T = 12\ \Omega$$

Now that the equivalent resistance for R_2 and R_3 has been calculated, the circuit can be redrawn as a series circuit as shown in Figure 1-26(B). The total resistance for the entire circuit may now be calculated as follows:

$$R_T = 8\ \Omega + 12\ \Omega$$

$$R_T = 20\ \Omega$$

The original circuit can be redrawn with a single resistor that represents the equivalent resistance of the entire circuit as shown in Figure 1-26(C).

$$I_T = \frac{60\ V}{20\ \Omega}$$

$$I_T = 3\ A$$

Summary

- A basic electric circuit consists of a source of electrical energy connected to a load. The load uses the energy and changes it to a useful form.
- A schematic diagram is a "picture" of a circuit that uses symbols to represent components. The space required to depict an electrical or electronic circuit is greatly reduced by the use of a schematic diagram.
- Ohm's law can be transposed to find one of the values in a circuit if the other two values are known.
- The same current flows through each part of a series circuit.
- The total resistance of a series circuit is equal to the sum of the individual resistances.
- The total voltage across a series circuit is equal to the sum of the individual voltage drops.
- The voltage drop across a resistor in a series circuit is proportional to the ohmic value of the resistor.
- The total power in a series circuit is equal to the sum of the individual power used by each circuit component.

CHAPTER 2

Meters

The functioning of conventional electrical measuring instruments has traditionally been based upon electromechanical principles. Their mechanical components usually work on dc electricity. Mechanical frequency meters are an exception. A meter that measures ac has a built-in rectifier to change the ac to dc, and resistors to correct the resistance for the various ranges.

Today, many meters utilize solid-state digital systems for operation; they are superior because they have no moving parts. These meters will work in any position, unlike mechanical meters, many of which must remain in one position in order to be read accurately.

When using any testing instrument, always consider the operator's personal safety first. Know the voltage levels and shock hazards related to all equipment to be tested, and make certain that the instrument used has been tested and correctly calibrated; this should be done at least once a year. To prevent damage to the test instrument, select a range (on meters with different ranges) that ensures less than full-scale deflection on readouts in the cases of needle or digital instruments, respectively.

d'Arsonval Meter Movement

The conventional ammeter is based on the principle that the magnetic effect of electricity in a current-carrying conductor will rotate when placed in a magnetic field, the same as an electric motor. The readings on an ammeter of this kind have been calibrated by comparison with a primary standard instrument. This calibration provides assurance that the practical magnetic instruments used in everyday electrical measurements indicate the same amount of current as the standard instrument.

In 1882, a Frenchman named Arsene d'Arsonval invented the galvanometer. The meter was basically a device that used a small moving coil suspended in a strong magnetic field as shown in Figure 2-1. When the current to be measured passes through the coil, a magnetic field is set up around the coil. A pointer is attached to the moving element, and therefore the pointer moves proportionately to the amount of current through the coil. The reading on the scale indicates directly how many amperes are passing through the instrument; that is, one particular amount of current causes the coil pointer to rotate to one particular place on the scale. A larger current causes the coil to rotate farther, while a smaller current causes the coil to rotate less. The early galvanometer was very accurate but could only measure very small currents. Over the following years, many improvements were made that extended the range of the meter and increased its ruggedness.

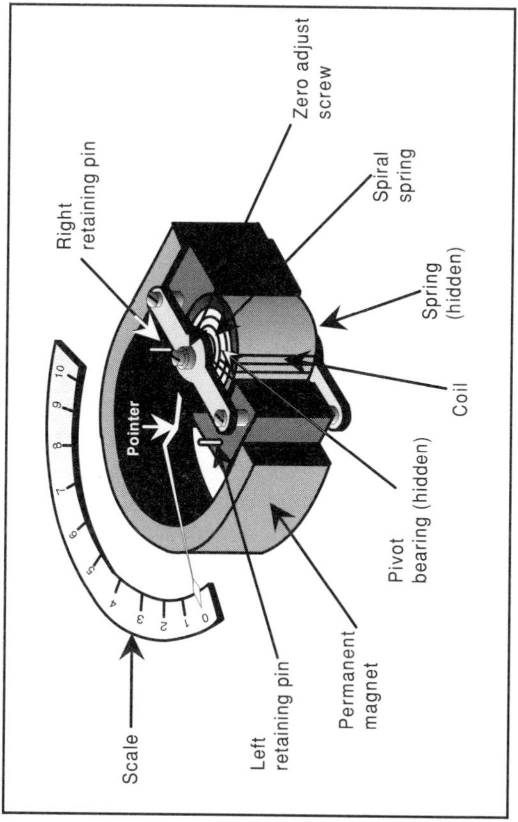

Figure 2-1: Major parts of a d'Arsonval meter.

The basic d'Arsonval meter operates on direct current (dc) only because alternating current (ac) continually reverses its direction and the meter coil will do likewise. However, by installing a rectifier that can convert ac to dc, the output is dc and the d'Arsonval meter can read it.

A moving-coil meter movement operates on the electromagnetic principle. In its simplest form, the moving-coil meter uses a coil of very fine wire wound on a light metallic frame. A permanent magnet surrounds the coil. The metallic frame is mounted on pivots to allow it and the coil to rotate freely between the poles of the permanent magnet. When current flows through the coil, it becomes magnetized, and the polarity of the coil is such that it is repelled by the field of the permanent magnet.

This will cause the coil frame to rotate on its pivots, and the distance it rotates is determined by the amount of current that flows through the coil. By attaching a pointer to the coil frame and adding a calibrated scale, the amount of current flowing through the meter can be measured.

Magnetic-Vane Meters

Another type of meter is called the magnetic-vane meter. It differs from the d'Arsonval meter in that the magnetic-vane meter contains a fixed coil and a movable

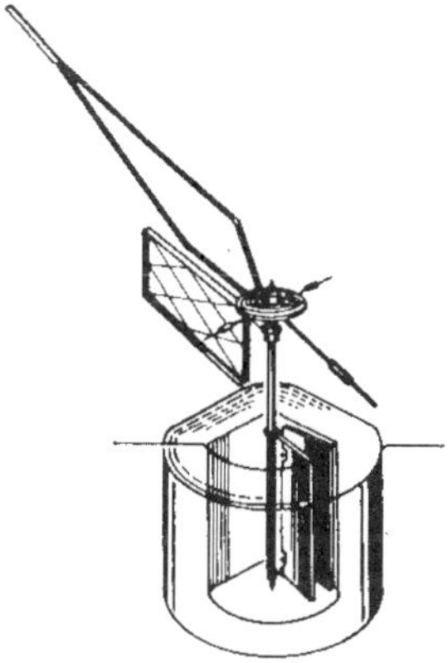

Figure 2-2: Working parts of a magnetic-vane meter.

iron vane, rather than the fixed magnet and moving coil used in the d'Arsonval meter.

Figure 2-2 shows a typical magnetic-vane meter. In this meter, two iron vanes are enclosed by a coil of wire. When there is current through the coil, the vanes are magnetized so that they repel each other regardless of whether the energizing current is direct or alternating. Usually one vane is fixed and the other is pivoted. The movable vane is provided with a pointer over a calibrated scale.

In recent years, digital-readout meters are rapidly taking the place of many of the needle-pointer-type meters. A digital meter displays the readings (voltage, amperage, ohms, etc.) in digits instead of using a meter movement (See Figure 2-3). Digital meters have several advantages over testing instruments with the conventional dial indicator. The greatest advantage is that the input impedance, or resistance, is higher. Dial or analog meters normally have a resistane of about 5000 Ω per volt. Therefore, on a 3-V full-scale range, the meter movement has a resistance of 15,000 Ω connected in series with it. Consequently, on a 600-V full-scale range, the meter movement has a resistance of 3 million Ω (3MΩ) connected in series with it.

Figure 2-3: Digital multimeters are available for taking voltage, amperage, and resistance readings.

On the other hand, digital meters commonly have an input impedance of 10 million Ω (10 MΩ), regardless of the range to which they are set to operate. The advantage of this high input impedance is that it does not interfere with a low-power circuit. On line voltage readings, the advantage is not very great. However, when working with the lower voltage control circuits, a low impedance meter will not only lead to inaccurate readings, but it may also alter the circuits themselves.

Another advantage of the digital meter is that it is generally easier for an inexperienced person to learn to read. Dial or analog meters can be used for the majority of testing and troubleshooting when working with electrical systems, but it usually takes more experience and practice for the technician to read the analog meters — especially on the lower values.

CHAPTER 3

Ammeters

An electrical circuit is a complete conductive path through which electrons may move. The movement of electrons in an electrical circuit is called current and is measured in amperes with an ammeter.

Most ammeters operate on a magnetic principle or the magnetic effect of electric current to produce the movement of the meter pointer on the scale or the digital readout on electronic instruments. A basic dc ammeter consists of a permanent magnet that supplies a magnetic flux or field, a delicately balanced coil that rotates in the field, a pointer, and a scale. This meter design is the d'Arsonval type that was discussed previously.

Most basic d'Arsonval type ammeters will measure only small amounts of current. The typical range is in microamperes (0.000001 A) or milliamperes (0.001 A). Very few ammeters in use can measure very much current without additional modifications. To increase the range of the ampere level, a shunt is used as shown in Figure 3-1 on the next page.

The meter is connected in parallel with the shunt. Shunts may be located inside the meter housing (internal shunt) and are generally used to measure currents up to 30 A. Ampere ratings over 30 A generate too much heat

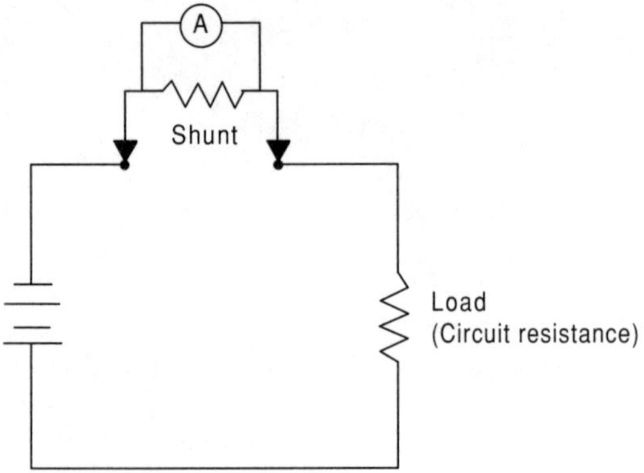

Figure 3-1: Ammeter shunt.

for internal shunts that could damage the meter. Consequently, other types of ammeters utilize external shunts with leads connecting the shunt to the meter. External-shunt ammeters are normally used to measure currents over 30 A. The use of a shunt allows the ammeter to derive current in amperes by actually measuring the voltage drop across the shunt.

To measure current above 10 mA, a shunt with an extremely low resistance is placed in series with the load and the meter is connected across the shunt to measure the resulting voltage drop proportional to current flow.

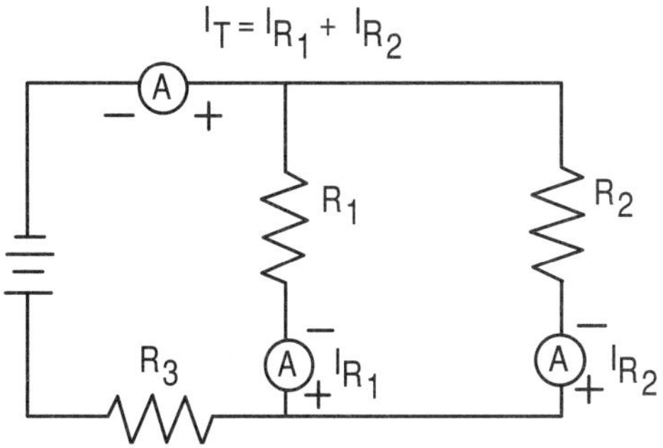

Figure 3-2: Ammeters connected at various points in a parallel dc circuit.

The shunt in Figure 3-2 has a relatively large wattage rating to enable it to carry a relatively large current.

CAUTION

Never connect an ammeter in parallel with a load. Because of the low resistance in the ammeter, a short-circuit can occur. Consequently, ammeters must be connected in series with the load. Furthermore, proper polarity must be observed in a dc circuit; that is, the negative lead of the meter must be connected to the negative side of the load, while the positive lead (+) must be connected to the positive side of the load. If the meter is connected with the polarities reversed, the meter coil will move in the opposite direction and often damage the meter.

As mentioned previously, alternating current can also be measured with a d'Arsonval-type meter if the alternating current is first changed to direct current by a rectifier. As the name implies, a rectifier is a device that rectifies, or converts ac to dc.

A thermocouple may also be used to adapt dc meters to the measuring of alternating current, but since thermocouple-equipped instruments are quite sensitive to overloads, their applications are usually limited to electronic circuits where overloads seldom occur.

Clamp-On Ammeters

Figure 3-3 shows a clamp-on ammeter that can be used to measure current in an electrical circuit while the circuit carries the full load. Disconnection of the conductors is not necessary to make the test.

Although the exact operating procedures will vary with the manufacturers, most operate as follows when measuring current:

Step 1. Make sure the battery-attachment case (for use when in the ohmmeter mode) is removed from the instrument.

Step 2. Release on pointer locks.

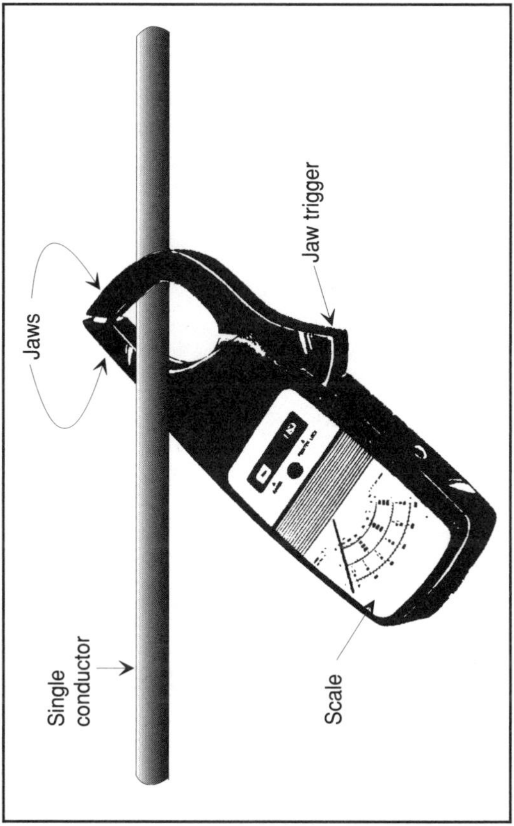

Figure 3-3: Typical clamp-on type ammeter.

Step 3. Turn the scale selector knob until the highest current range appears in the scale window.

Step 4. Press the trigger button to open the jaws of the clamp before encircling one of the conductors under test with the transformer jaws.

CAUTION

Never encircle two or more conductors as shown in Figure 3-4(A); only encircle one conductor as shown in Figure 3-4(B).

Step 5. Release finger pressure on the trigger slowly to allow the jaws to close around the conductor and keep an eye on the scale while doing so. If the pointer jumps abruptly to the upper range of the scale before the jaws are completely closed, the current is probably too high for the scale used. Should this happen, remove the jaws immediately from the conductor and use either a higher scale or a range-extender attachment as discussed in the next section. If the pointer deflects normally, close the jaws completely and take the reading from the scale.

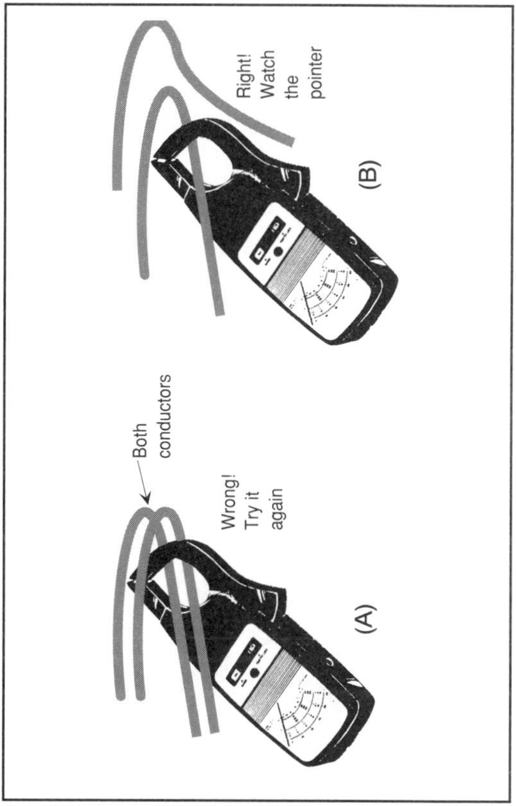

Figure 3-4: Right and wrong way to take current readings with a clamp-on ammeter.

Although the clamp-on ammeter is very convenient to use, the operator may possibly be close to high voltage that is potentially dangerous. Therefore, special precautions must be taken to prevent accidental contact with live conductors and parts. When using this type of instrument, the current-carrying conductor should be centered inside the iron-core clamp for greatest accuracy.

Extending the Range of Ammeters

The range of an ac ammeter can be extended by using a range extender as shown in Figure 3-5. This device permits taking a measurement of higher current beyond

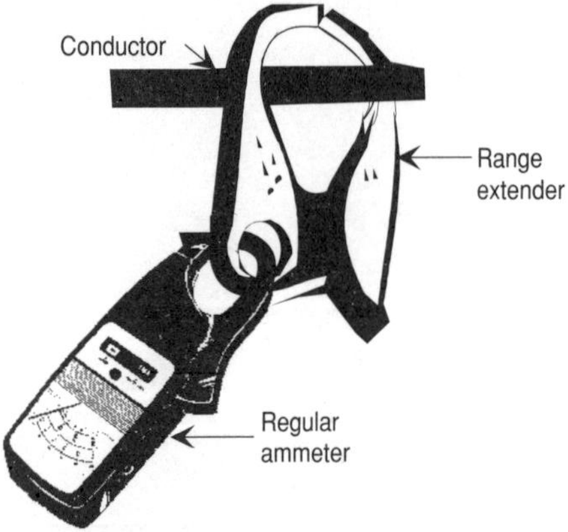

Figure 3-5: Clamp-on transformer extends ammeter range.

the range of the regular clamp-on ammeter. The model shown extends the current range ten times to allow an actual current reading of 1000 A on a 0- to 100-A meter scale. To illustrate its use, if the scale shows a reading of, say, 42 A, the actual current (using the range extender) would be 420 A because $42 \times 10 = 420$.

Current Multipliers

Sometimes it is desirable to use a current multiplier with a clamp-on ammeter, such as the one shown in Figure 3-6. This device allows current measurement on low-current equipment since the load current shown has been multiplied either two, five, or ten times; that is, if

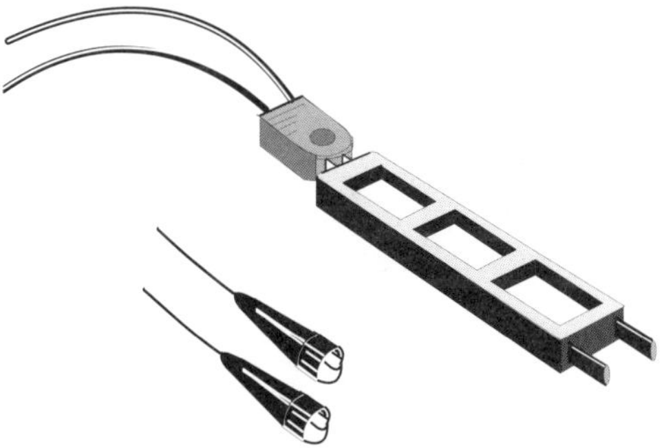

Figure 3-6: Current multiplier in use.

the meter scale shows a reading of, say, 62 A and the 10× multiplier was used, the actual load current would be:

$$\frac{62}{10} = 6.2 \textit{ amperes}$$

If a current-multiplier attachment is not available, the same effect may be had by wrapping turns of wire around the jaws of the ammeter as shown in Figure 3-7. In this illustration, the wires are wrapped around the jaws to multiply the current three times. If an attachment is available, and the device or apparatus to be tested has a plug-and-cord assembly, the load may be tested as shown in Figure 3-8.

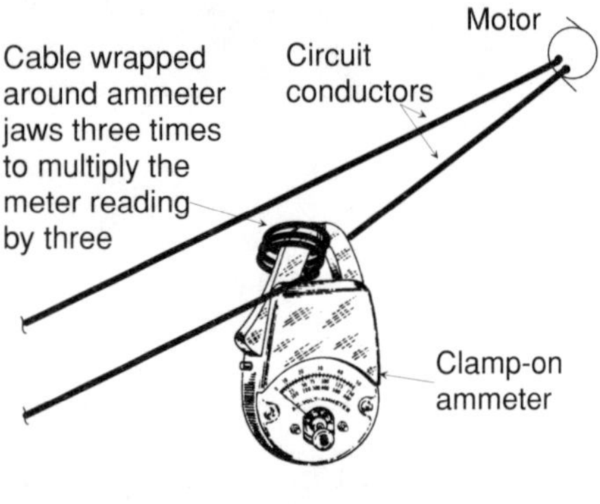

Figure 3-7: Wrapping turns of wire around meter jaws will increase the current reading.

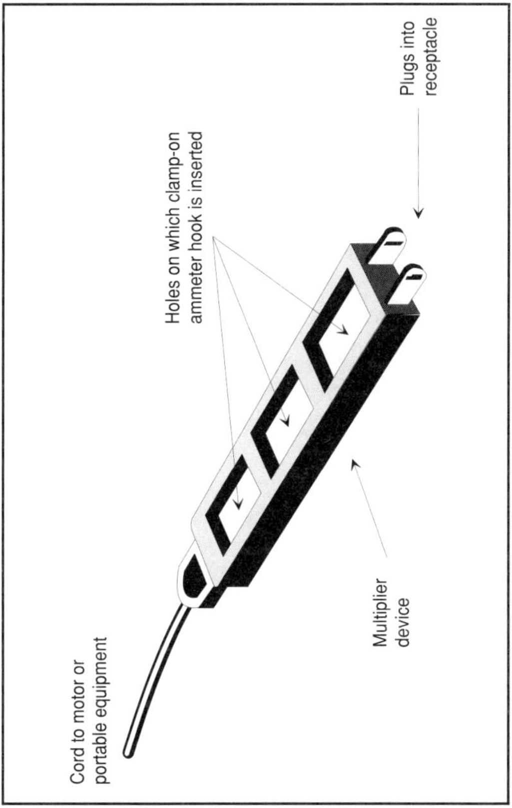

Figure 3-8: Ammeter multiplying device for use on plug-in equipment.

Precautions

When using clamp-on ammeters, care must be taken to obtain accurate readings. Some items to be considered include:

1. Make certain that the frequency of the circuit under test is within the range of the instrument. Most are calibrated at 70 hertz (Hz).

2. Take care that stray magnetic fields do not affect the current reading; that is, arrange the meter leads as far away as possible from the conductors under test.

Try to take current readings in a control panel at a location remote from magnetic relays that might influence the accuracy of the reading.

Avoid taking current readings on conductors at a point close to a transformer.

3. When current readings are taken on high-voltage conductors, always use a hot-line extension pole specifically designed for use with a high-voltage clamp-on ammeter.

Ammeters are very useful instruments for testing and troubleshooting electrical circuits. The ammeter, when used in conjunction with a voltmeter, can also be used to find both the resistance and volt-amperes (VA) in a circuit.

For example, assume that the resistance of an electric lamp is to be determined. The method to use is illustrated in Figure 3-9. First, a dc source (battery) is connected to the lamp element; a voltmeter is then connected across the load, and an ammeter is connected in series with the circuit. The resistance of the element can then be found by applying Ohm's law:

$$R = \frac{E}{I}$$

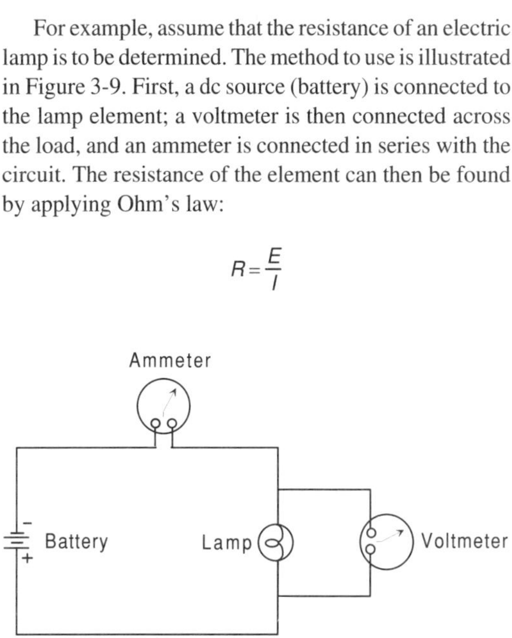

Figure 3-9: Ammeter and voltmeter used to determine load resistance.

where R = resistance in ohms

E = voltage in volts

I = current in amperes

Therefore, if the voltage reading across the lamp element is 6 volts and the ammeter reading is 0.3 A, by substituting these values in the equation, we have:

$$R = \frac{6}{0.3} = 20\Omega$$

You may wonder why an ohmmeter was not used to merely measure the resistance in the lamp, rather than going to the trouble of connecting the two meters in question. A "cold" lamp has a different resistance from a "hot" or energized lamp. Therefore, the true resistance of an incandescent lamp cannot be determined by the use of an ohmmeter.

Practical Applications of Ammeters

The ammeter is quite handy for troubleshooting various electrical components by indicating a change in electrical value. Many examples, including numerous troubleshooting charts, may be found throughout this book. In the meantime, let's take a couple of simple examples to demonstrate the usefulness of an ammeter.

Sometimes it is desirable to know the approximate load on a three-phase motor. This may be determined by taking a current reading while the motor is operating. All three phases should be checked, one at a time.

If the voltage is approximately that of the rated voltage of the motor, and the ammeter shows that the motor is drawing a current close to the nameplate rating, it may be reasonably assumed that the motor is carrying a full load. If the ampere reading is much less, the motor is not carrying a full load. If the current is much higher than the nameplate rating after the motor has come up to full speed, it may be assumed that the motor is overloaded.

To determine if, say, an electric baseboard heater is operating properly, examine the nameplate to find the heater's characteristics. Let's assume that the nameplate indicates a 1000-W single-phase, two-wire heating element operating at 240 A. If an ammeter reading is taken while the heater is operating and the reading shows approximately 4 A of current, it may be assumed that the heater is working as it should, because:

$$I = \frac{P}{E} \text{ or } \frac{1000}{240} = 4.16 \text{ amperes}$$

If the ampere reading is much different from 4 amperes, some fault in either the circuit or the heater is present.

Recording Ammeters

When it is desired to have a continuous and/or permanent record of the current in a given electrical circuit, a graphic or recording instrument may be used. Such an instrument has a meter element similar to the conventional indicating ammeters, but, in addition, it is equipped with a pen or other marking device so that a curve is drawn as current changes occur. The marking device on the recording instrument replaces the pointer on scale-indicating meters, and the marking device traces a line on a chart representing the value of the amperes that the instrument is measuring (see Figure 3-10).

The charts are usually either circular or in strip form, depending on the type of instrument.

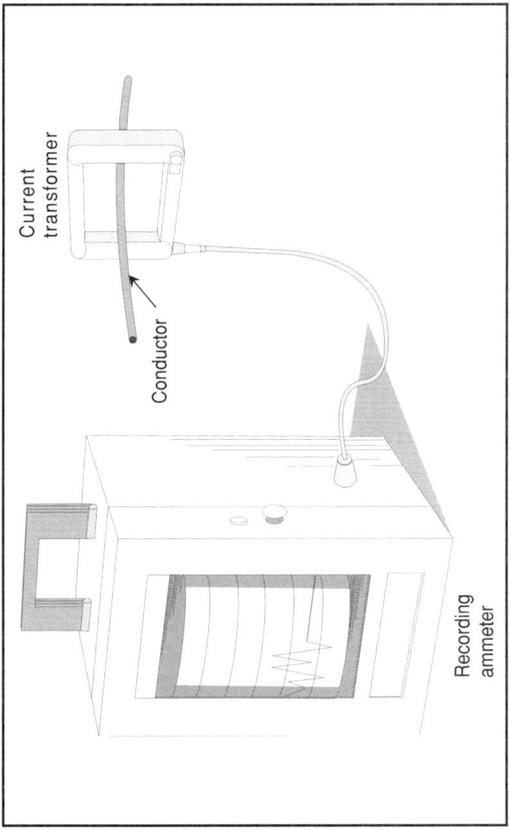

Figure 3-10: Recording ammeter.

CHAPTER 4

Voltmeters

The unit of electromotive force (emf) is the volt (V). One volt is a form of pressure that, if steadily applied to an electrical circuit having a resistance of 1 Ω, produces a current of 1 A.

Voltmeters are used to accurately measure the pressure or voltage in various electrical circuits. Very low voltage values are measured in millivolts (a volt = 1000 mV) by a millivoltmeter with low resistance.

A voltmeter should be connected across the terminals at the place the voltage is to be measured, as shown in Figure 4-1. A voltmeter should never be connected across a circuit having a voltage higher than the rating of the instrument; that precaution must be observed particularly in the case of measurements with a millivoltmeter.

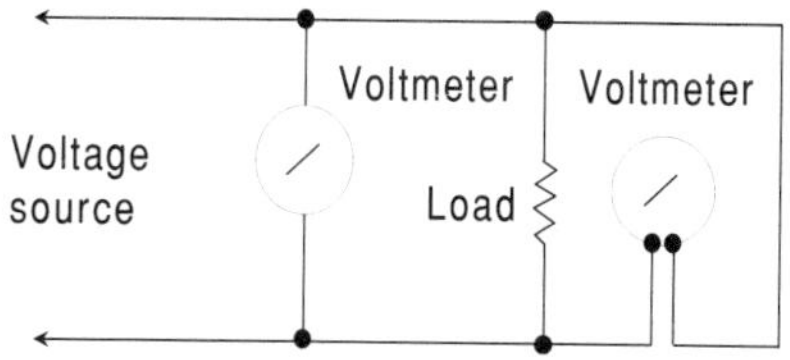

Figure 4-1: Voltmeter connection methods.

Several types of voltmeters are in common use. For example, the basic pointer-type ammeter can also be used to measure voltage. The meter coil has a fixed resistance, and, therefore, when current flows through the coil, a voltage drop will be developed across this resistance. According to Ohm's law, the voltage drop will be directly proportional to the amount of current flowing through the coil. Also, the amount of current flowing through the coil is directly proportional to the amount of voltage applied to it. Therefore, by calibrating the meter scale in units of voltage instead of current, the voltage in different parts of the circuits can be measured.

Since the basic current meter movement has a low coil resistance and low current handling capabilities, its use as a voltmeter is limited. In fact, the maximum voltage that could be measured with a 1 milliamp meter movement is 1 V (see Figure 4-2).

The voltage range of the meter movement, however, can be extended by adding a resistor, called a multiplier resistor, in series. The value of this resistor must be such that, when added to the meter coil resistance, the total resistance limits the current to the full-scale current rating of the meter for any applied voltage (see Figure 4-3 on page 90).

When connecting a voltmeter to a dc circuit, always observe the proper polarity. The negative lead of the meter must be connected to the negative terminal of the dc source, and the positive lead to the positive terminal.

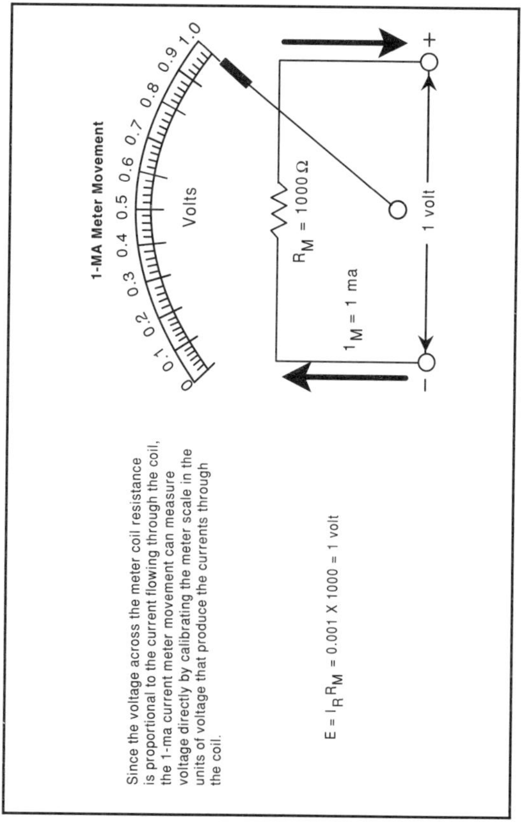

Figure 4-2: Pointer movement of a 1 mA meter.

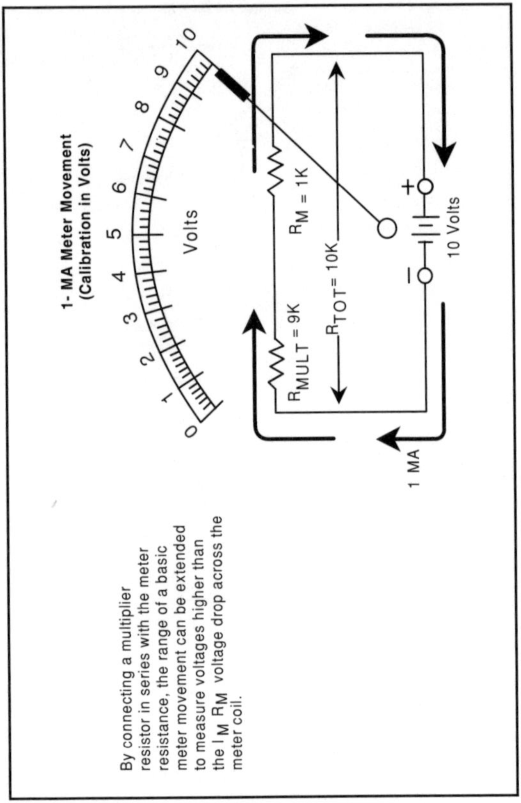

Figure 4-3: Adding multiplier resistor to a voltmeter.

If the leads are connected to opposite terminals, the needle will move in the reverse direction. Since the voltage constantly reverses polarity in an ac circuit, there is no need to observe polarity when connecting voltage designed for use on ac circuits.

Many portable voltmeters are designed with two or more voltage ranges that can be read on a common scale, such as 0 to 150 V, 0 to 300 V, and 0 to 600 V. The change from one meter range to another is accomplished by the selection of internal shunts, resistors, or other auxiliary components by means of a switch or plug-in terminals. When using a multirange voltmeter, it is best to select a higher range than needed to assure that no damage will occur to the instrument. Then, if the initial reading indicates that a lower scale is needed to obtain a more accurate reading, the meter can be switched or otherwise adjusted to the next lowest range.

To illustrate the use of a voltmeter at a conventional residential wall receptacle, turn the selector switch to 0- to 300-V range, since the highest voltage that will be present in any single-family dwelling is 240 V. The test leads of the meter are then placed in the receptacle slots to read the voltage as shown in Figure 4-4 on the next page. Usually, such a test will show between 110 and 120 V on the meter. Since the selector switch is in the 0- to 300-V position, the meter pointer should indicate a value somewhat below half-scale. The voltage (110 to 120 V) is less than 150 V, so the selector is then turned

Figure 4-4: Checking voltage at a 125-V duplex receptacle.

to the 0- to 150-V range. The meter should then give a reading within 1 V of the exact voltage at the receptacle.

One of the reasons for using various ranges of voltmeters is that the greatest accuracy is obtained on the upper half of the scale. Therefore, if a single 0- to 600-V range were used, lower voltages should be harder to read and meter accuracy would be less.

Some multirange voltmeters have only one range of values marked on the scale, and the scale reading must be multiplied by the range-switch setting to obtain the correct voltage. For example, the voltmeter scale in Figure 4-5 has only one set of voltage values. Consequently, if the range (selector) switch is set to 1 V

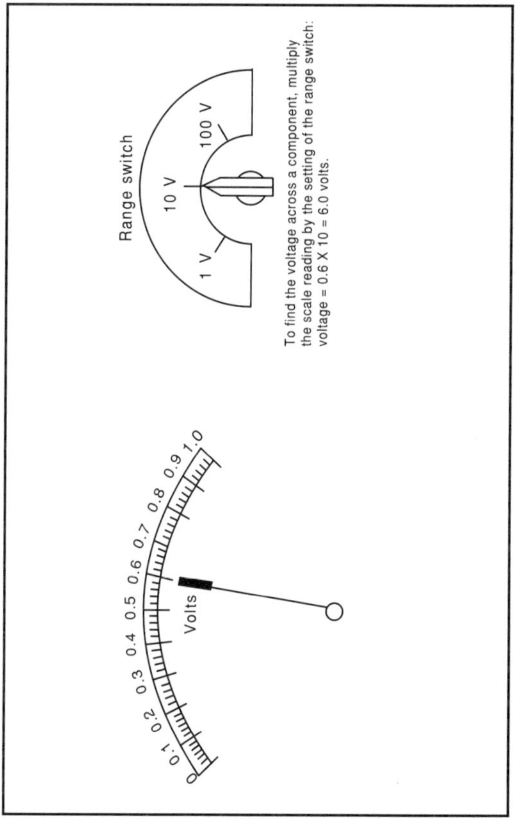

Figure 4-5: Multirange, one-scale voltmeter.

(sometimes marked 1×), and the meter shows a reading of 0.6, the voltage equals 0.6 V. However, if the selector switch is moved to 10 V, and another voltage test indicates 0.6, this is multiplied by 10 ($0.6 \times 10 = 6.0$) for a voltage rating of 6 V. If the selector switch is moved to 100 V, the value indicated on the scale must be multiplied by 100 to obtain the correct voltage value.

Other voltmeters have different ranges on the scale for each setting of the range switch (see Figure 4-6). In this case, be sure to read the set of values that corresponds to the position of the range switch.

Solid-state digital voltmeters are usually autoranging; that is, they do not have a selector switch for adjustment. The internal construction of the meter itself will select the proper resistance for the voltage being detected.

Voltmeter Applications

Voltmeters can be used for other electrical tests, such as troubleshooting circuits, circuit tracing, and measuring low resistance. For example, a common cause of electrical problems is low equipment voltage. This problem usually occurs because of one or more of the following reasons:

- Undersized conductors
- Overloaded circuits
- Transformer taps set too low

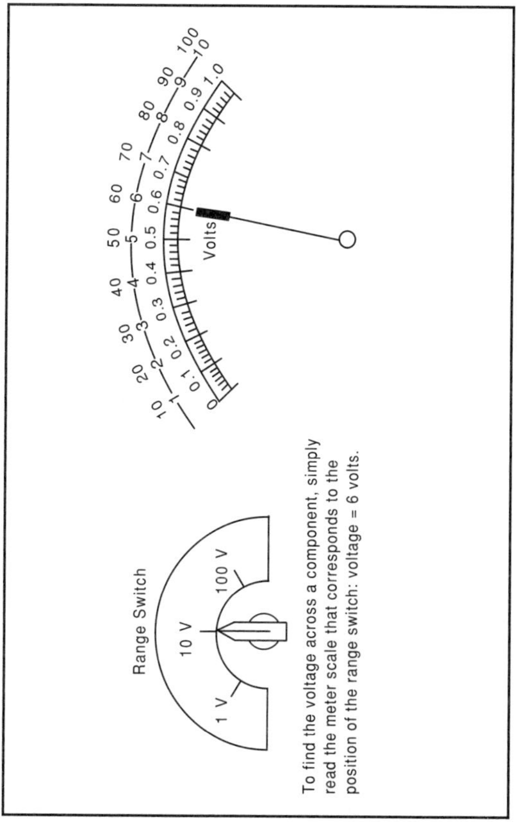

Figure 4-6: Multirange, multiscale voltmeter.

Low-Voltage Test

When making a low-voltage test, first take a reading at the main switch or service entrance. If, for example, the main service is 120/240, single-phase, 3-wire, the voltage reading between phases (ungrounded conductors) should be between 230 and 240 V. If the reading is much lower than this, the fault lies with the utility company supplying the power, and they should be notified to correct the problem. However, if the reading at the main service switch is between 230 and 240 V, the next procedure is to check the voltage reading at various outlets throughout the system.

When a low-voltage problem is found on a circuit, leave the voltmeter terminals connected across the line and begin disconnecting all loads, one at a time, that are connected to the circuit. If the problem is corrected after several of the loads have been disconnected, the circuit is probably overloaded. Steps should be taken to reduce the load on the circuits or else increase the wire size to accommodate the load. As mentioned previously, loose connections can also cause low voltage; to check for this, the entire circuit should be de-energized, and each terminal in all disconnect switches, motor starters, and so on, should be checked for loose connections. A charred or blackened terminal screw is one sign to look for in the various components of the system.

Ground Fault

Another problem is a ground fault in a circuit. For example, assume that a three-phase, three-wire, 240 V, delta-connected service is used for power in a small industrial plant. The service equipment is installed as shown in Figure 4-7 on the next page. Under proper operating conditions, the voltmeter should read 240 V between phases (A and B, B and C, A and C), and approximately 150 V between each phase to ground.

However, if on checking with the voltmeter, two of the readings show a voltage of 230 V between phase and ground and the other reading is only 50 V between phase and ground, then the phase with the lowest reading (50 V) has a partial ground or ground fault.

The sequence for correcting the ground fault is as follows:

Step 1. Connect one of the voltmeter terminals to the grounded main-switch housing and the other to the phase terminal that indicated the ground fault.

Step 2. Disconnect switch A (Figure 4-7) and check the voltmeter reading. If no change is indicated, disconnect switch B, switch C, and so on, until the voltmeter shows a change;

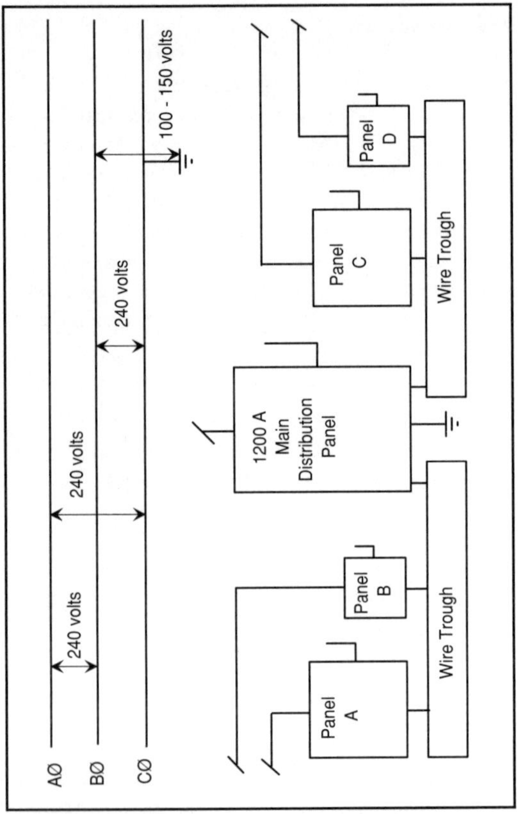

Figure 4-7: Diagram of a small industrial electric service.

that is, a reading of approximately 150 V from phase to ground.

Step 3. Assuming that the voltmeter indicates this reading when switch D is thrown to the OFF position, we then know that the ground fault is located somewhere on this circuit.

Step 4. Switch D disconnects the 400-ampere circuit feeding eight 15-horsepower (hp) motors and connected as shown in Figure 4-8 on the next page. One voltmeter lead is connected to the grounded housing of switch D and the other lead to one of the phase terminals. The switch is then thrown to the ON position. Check each phase terminal until the one with the ground fault is located.

Step 5. Then the motors, one at a time, are disconnected from the circuit until the one causing the trouble is found; that is, when the motor or motor circuit having the ground fault is disconnected, the voltmeter will show a normal voltage of approximately 150 V from phase to ground.

Step 6. The faulty motor or motor circuit can then be corrected according to standard maintenance procedures.

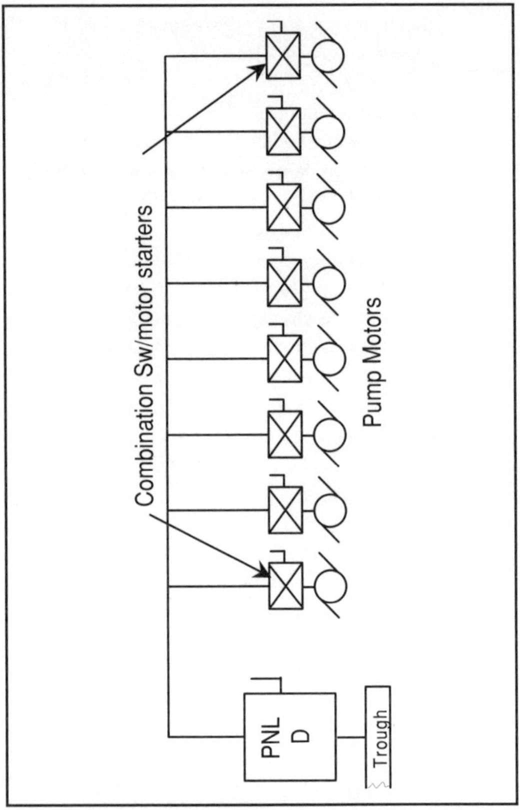

Figure 4-8: Wiring diagram for eight 15 hp pump motors fed from a 400-A safety switch.

When testing electrical circuits with a voltmeter, it is usually best to begin at the main service disconnect. First, test the voltage on the line side to see if the line is "hot" from the outside service wires; if it is, then test the fuses. The fuses may be checked by testing across diagonally from the line to the load side as shown in Figure 4-9.

Circuit Tracing

A dc voltmeter can be used for tracing circuits when one lead of the meter is connected to a 6-V battery

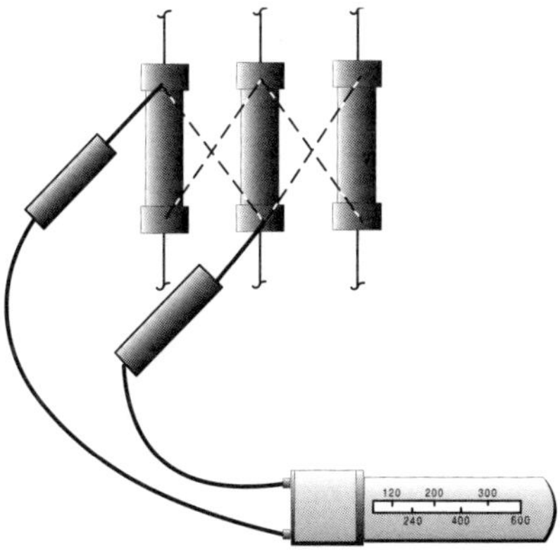

Figure 4-9: Correct method for testing fuses with a voltmeter.

terminal; a jumper wire from the other battery terminal is then connected to a common point, such as to the rigid conduit containing the wires. One end of the wire that is to be traced is connected to the common point or conduit, and the other ends of the wires are left open. The other lead from the voltmeter is then connected to the open ends of the wires in turn. When a reading is indicated on the voltmeter, the circuit is completed and this is the other end of the wire in question. This process may be repeated to locate any number of wires in the conduit or system. In this situation, the voltmeter becomes a continuity tester.

Measuring Low Resistance

A dc voltmeter can also be used to measure a low resistance in a circuit when used in conjunction with a known resistance. For example, an unknown resistance is connected to a circuit as shown in Figure 4-10. To measure the resistance, another resistor of known value (say, 0.25 Ω) is connected in series with the unknown resistor and then both are connected to a 6-V battery (Figure 4-11). The same current will pass through both resistors (and the voltage drop across them is directly proportional to their resistance) and then across the known resistance as shown in Figure 4-12. The two readings of the voltmeter are directly proportional to resistances R1 and R2, respectively, and the unknown resistance may be found by dividing the voltmeter read-

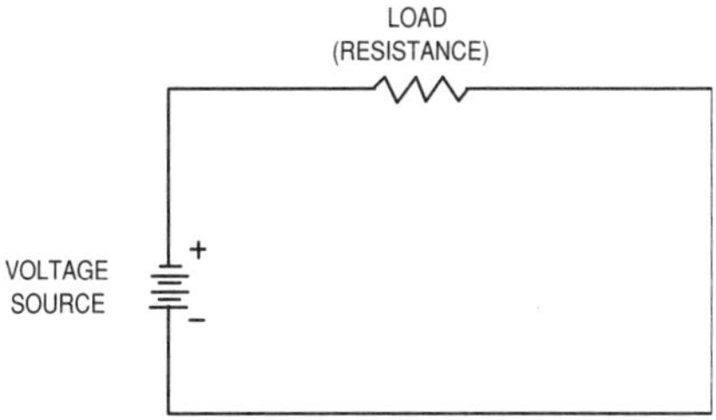

Figure 4-10: Unknown resistance in an electrical circuit.

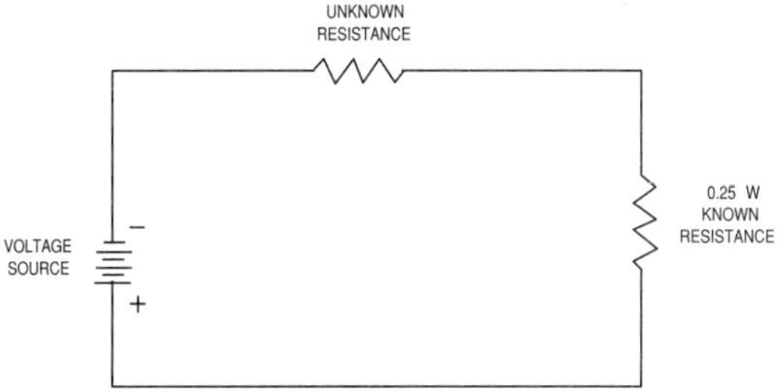

Figure 4-11: A resistor of known value connected to the circuit in Figure 4-10.

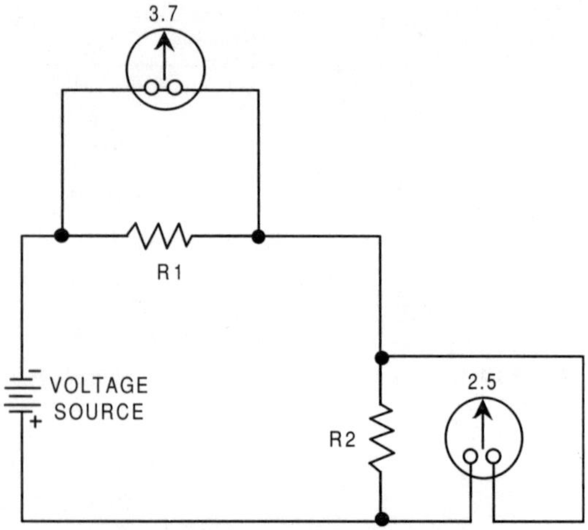

Figure 4-12: Voltmeter connections to determine an unknown resistance.

ing across the known resistor times the known resistance into the voltmeter reading across the unknown resistor. This calculation will give the unknown resistance.

Thus, Figure 4-12 shows that the voltmeter reading across the unknown resistance is 3.7 V and 2.5 V across the known resistance of 0.25 Ω. Substituting these values in the equation, we have:

$$\frac{3.7}{2.5} \times 0.25 = 0.37\ \Omega$$

Therefore, the resistance of the unknown resistor is 0.37 Ω.

If the internal resistance of the voltmeter is known, the value of an unknown resistance can be found using only the voltmeter. For example, the internal resistance of a voltmeter is known to be 16,000 Ω and it is connected across an electrical circuit as shown in Figure 4-13. If the meter reading is 240 V, the resistance of an unknown resistor can be found by connecting it in series with the voltmeter and then reading the voltage again. If the voltage reading drops to 238 V, the value of the resistor can be found by the following equation:

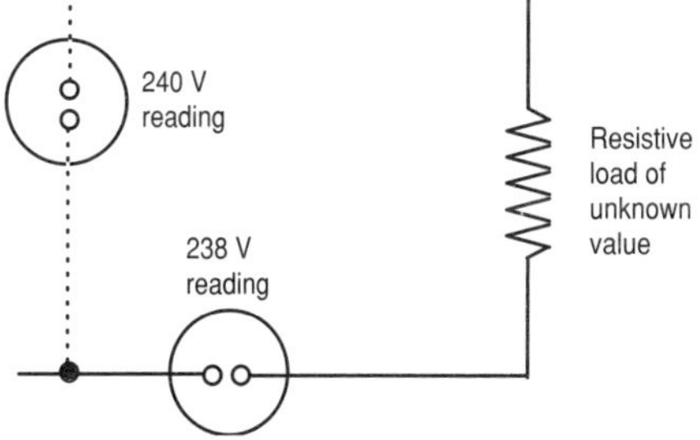

Figure 4-13: Connect voltmeter with a known resistance in series with a load resistance.

$$16{,}000\Omega \times \frac{240}{238} - 1 = 134.4\ \Omega$$

where: 16,000 = internal resistance of voltmeter in ohms

240 = reading across circuit in V

238 = reading across unknown resistor in V

–1 = a constant factor

The voltmeter is a very useful instrument for all personnel working with electricity, but its capabilities must be thoroughly understood in order to obtain its full usefulness. Proper care and maintenance are also necessary to assure accurate readings and reliability. Figure 4-14 shows two popular kinds of voltmeters in use today. The meter A in Figure 4-14 is a combination volt-ohm-ammeter with a conventional swinging pointer to indicate the reading; meter B in Figure 4-14 works on the "ting" principle — similar to an air gauge — and gives only approximate voltage readings.

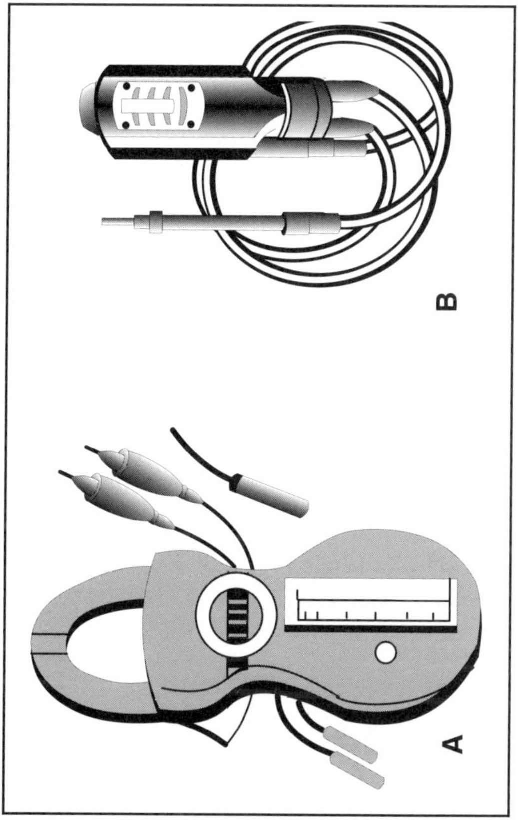

Figure 4-14: Some common types of voltmeters.

CHAPTER 5

Megohmmeters

A typical megohmmeter (megger) is composed of a hand-driven or motor-powered ac generator and/or a transformer with voltage rectified to 100, 250, 500, and 1000 V dc, a cross-coil movement with 0 to 20,000 ohms (Ω) and 0 to 1000 megohms (MΩ) scales, a carrying case, and test leads. The megger (Figure 5-1) is used to measure the resistance in megohms to the flow of current through and/or over the surface of electrical equipment insulation. The test results are used to detect the presence of dirt, moisture, and insulation deterioration. The instrument also typically measures resistance up to 20,000 W.

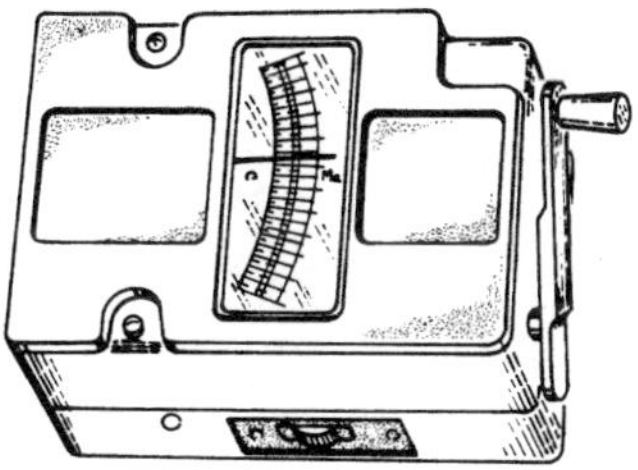

Figure 5-1: Typical hand-driven megger.

The test set and the sample to which it is connected are sources of high-voltage electrical energy, and all persons making or assisting in the tests must use all practical safety precautions to prevent contact with energized parts of the test equipment and associated circuits. Persons actually engaged in the test must stand clear of all parts of the complete high-voltage circuit unless the set is deenergized and all parts of the test circuit are grounded. If the set is properly operated and all grounds are correctly made, no rubber gloves are necessary. As a routine safety procedure, however, some applications require the use of rubber gloves in making connections to the high-voltage terminals and in manipulating the controls.

The instruction manuals accompanying the megger contain detailed instructions about preparing for tests and connecting the megger to various types of equipment. Figures 5-2, 5-3, and 5-4 give some practical points on hookups of megger instruments.

Figure 5-2 shows that ac motors and starting equipment can be megger-tested by connecting one side of the megger to the motor side of the main switch and the second test connection to a clip on the motor housing. If an insulation weakness is indicated by this test, the motor and starter should be checked separately.

Figure 5-3 shows the connection for testing low-voltage power cable. After both ends of the cable have been disconnected, the conductors are tested, one at a time, by

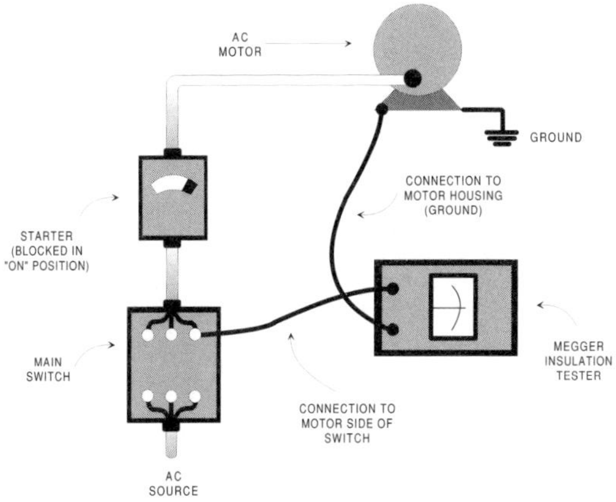

Figure 5-2: Megger connections for testing ac motors and starting equipment.

connecting one of the megger leads to the conductor under test and connecting the remaining conductors (within the cable) to ground and then to the other (ground) test lead. Similarly, other insulation resistance, such as between conductor and outside protective sheath and between conductors, can be measured.

Lighting and distribution transformers are tested by first making sure that switches or circuit breakers on both the primary and secondary sides are open. The high-voltage winding to ground and low-voltage winding to

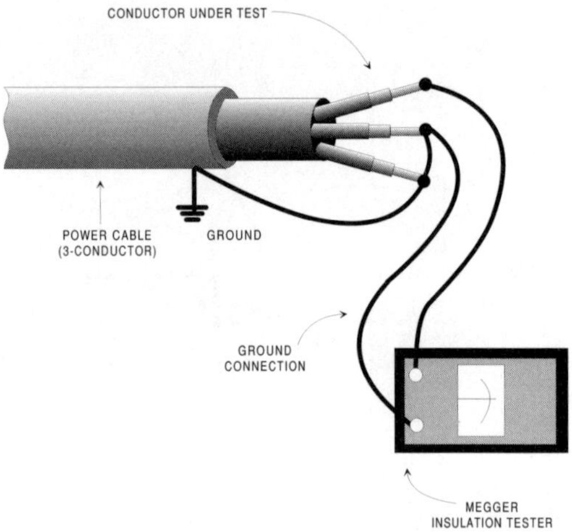

Figure 5-3: Megger connection for testing low-voltage power cable.

ground are consecutively checked by separate tests. The resistance between the two is then checked with neither of them grounded.

If an insulation weakness is shown by the test in an oil-filled transformer, the dielectric strength of the oil may be tested, if the reading is low, the transformer should be filled to remove moisture and sludge. If the windings still give a poor reading when tested, they probably need drying out. See Figure 5-4.

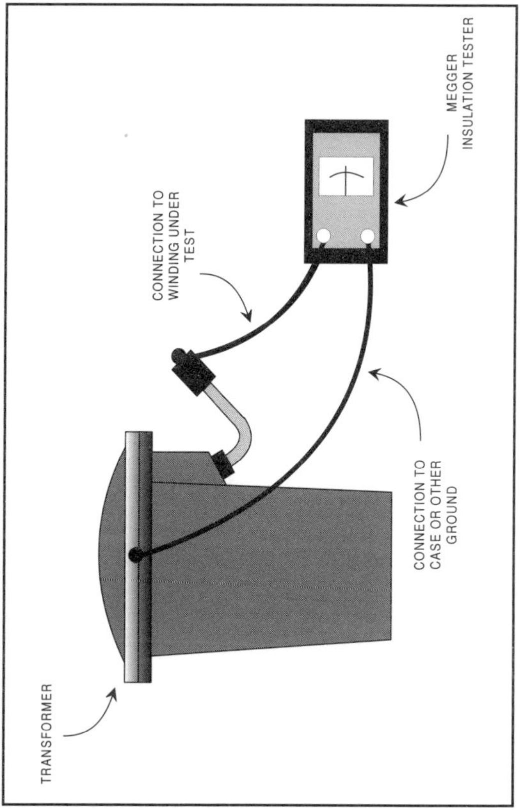

Figure 5-4: Testing lighting and distribution transformers with a megohmmeter.

Insulation Resistance Testing

To prepare for an insulation resistance test, first take the equipment or circuits to be tested out of service. Check between the equipment terminals and ground using the megger voltage ranges to be sure there is no voltage present. If possible, disconnect all leads to the unit being tested. When a motor or circuit is not completely isolated, make sure you are aware of all the components that will be tested when the megger is connected. Should an interconnected circuit be overlooked, the megger reading may be lower than expected.

The testing of wiring can be performed on all types of systems if two rules are kept in mind:

- Be sure all wiring is deenergized.
- Know what wiring is included in the test and make a record card of it.

When a distribution is present, check the entire system to ground by attaching one megger lead to the dead post of the open main power switch and the other lead to a grounded conduit or grounded metal housing.

Individual circuits are tested to ground by opening distribution panel switches, fuses, or circuit breakers and testing each circuit in turn.

Multiconductor cables may be tested in several ways. For instance, measurement of insulation resistance can be made between the wire and lead sheath. Various other measurements can be made, such as wire to ground, wire to wire, wire to braid, and wire to sheath.

Keep in mind that when testing wiring that is connected to any panelboard or equipment there may be appreciable leakage between terminals, which will show in tests as lowered insulation resistance. If previous test and record cards were made with the panels connected, continue any future test in the same manner.

WARNING!

Only qualified personnel should make tests with a megger, and then only after obtaining permission from their supervisor. Megger tests can be dangerous to personnel and harmful to equipment and wiring systems if improperly handled.

Testing DC Motors and Generators

To test dc motors and generators, disconnect the apparatus from any power source or load and attach the negative test lead of the megger to the machine ground and the positive lead to the brush rigging. Measuring the insulation resistance in this manner indicates the overall resistance of all components of the unit.

To measure the insulation resistance of the field or armature alone, either remove the brushes or lift them free of the commutator ring and support them by means of a suitable insulator. Connect one megger test lead to the frame ground and the other to one of the brushes.

Insulation resistance of the field alone will then be indicated as shown in Figure 5-5. With the brushes still removed from the commutator ring, connect one of the megger test leads to one of the segments of the commutator and the other to the frame ground. The insulating resistance of the armature alone will then be indicated. This test may be repeated for all segments of the commutator.

Testing AC Motors

To test ac motors, first disconnect the motor from the line, either by use of the switch or by disconnecting the

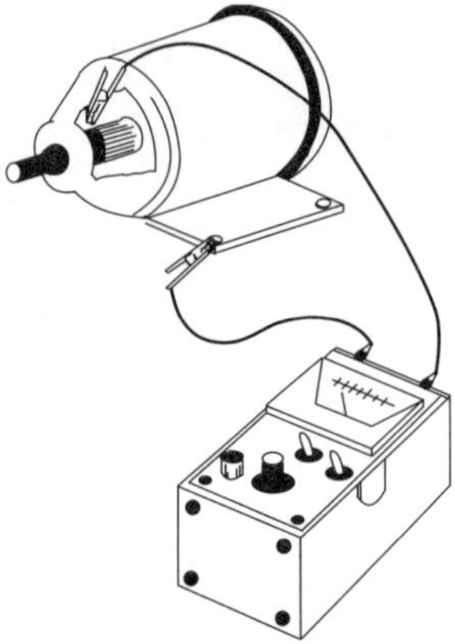

Figure 5-5: Megger connections for testing dc motors and generators.

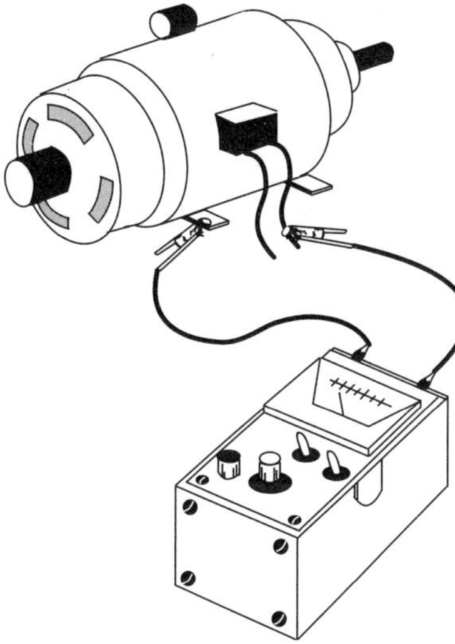

Figure 5-6: Method of testing an ac motor.

wiring at the motor terminals. If the switch is used, remember that the insulation resistance of the connecting wire, switch panel, and contacts will all be measured at the same time. Connect the positive megger lead to one of the motor lines and the negative test lead to the frame of the motor, as shown in Figure 5-6. If insulation resistance minimums have been established, the reading can be checked against them.

Testing Circuit Breakers

Circuit breakers may be tested with the megger by first disconnecting the circuit breaker from the line and connect the megger black lead to the frame or ground. Check the insulation resistance of each terminal to ground by connecting the red (positive) lead to each terminal in turn and making the measurements. Next open the breaker and measure the insulation resistance between terminals by putting one lead on one terminal and the other on the second for a two-terminal breaker; for a three-pole breaker, check between poles 1 and 2, 2 and 3, and 1 and 3, in turn.

If the breakers are of the oil type, dirty oil may cause leakage currents between the breaker contacts even when open. In dry types of breakers, dirt or grease in the mounting may cause leakage. The measurement of insulation resistance can be made between either circuit breaker terminal to ground and should also be made between terminals.

Testing Safety Switches and Switchgear

Switches with safe insulation are a vital part of any electrical installation and so deserve careful attention. Switches should be completely disconnected from the line and relay wiring before testing. When manual switches are being tested, measure the insulation resistance from ground to terminals and between terminals. When testing electrically operated switches, check the insulation resistance of the coil or coils and contacts. For coils, connect one megger lead to one of the coil leads and the other to ground. Next, test between coil lead and core iron or solenoid element.

If relays are involved, measure insulation resistance with one lead connected to the relay plate or contact and the other test lead connected to the coil, core, or solenoid contact.

Although the suggested tests may appear to be excessive in view of the usually superior performance of most switches over long periods of time, these recommendations are made with the knowledge that one inoperative closed switch can literally wreck a whole plant system. Thorough testing can help to ensure against such an occurrence.

Ground Resistance

Figures 5-7 and 5-8 show two basic methods for testing the resistance of earth. The direct or two-terminal test (Figure 5-7 on the next page) consists of connecting terminals P_1 and C_1 of the megger to the ground under test, and terminals P_2 and C_2 to an all-metallic water-pipe system. If the water-pipe system covers a large area, its resistance should only be a fraction of an ohm and,

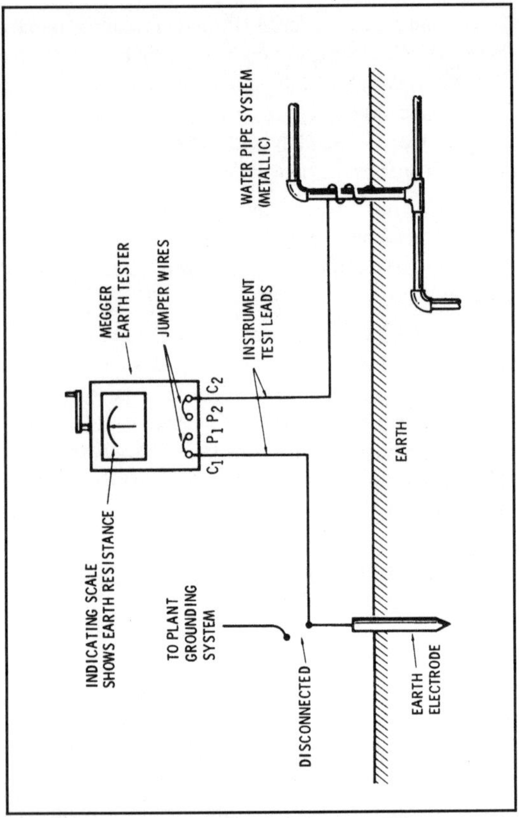

Figure 5-7: Direct method of earth-resistance testing.

therefore, the meter reading will be that of the ground or electrode under test.

The direct method is the simplest way to make an earth-resistance test. With this method, resistance of two electrodes in series is measured — the driven ground rod and the water system. But there are some important limitations:

- The water-pipe system must cover enough area to have a negligible resistance.
- The water-pipe system must be metallic throughout, without any insulating couplings or flanges.
- The driven ground rod or electrode under test must be far enough away from the water-pipe system —about 10 times the radius of the electrode — to be outside its sphere of influence.

The fall-of-potential method (Figure 5-8 on the next page) consists of using terminals C_1, C_2, and P_2 of the instrument. The P_1 and C_1 terminals on the instrument are jumped (connected) together and connected to the electrode or ground under test. The driven reference rod connected to terminal C_2 should be placed as far from the ground electrode under test as practical.

The potential-reference rod connected to terminal P_2 is then driven into the ground at a number of points roughly on a straight line between the ground under test

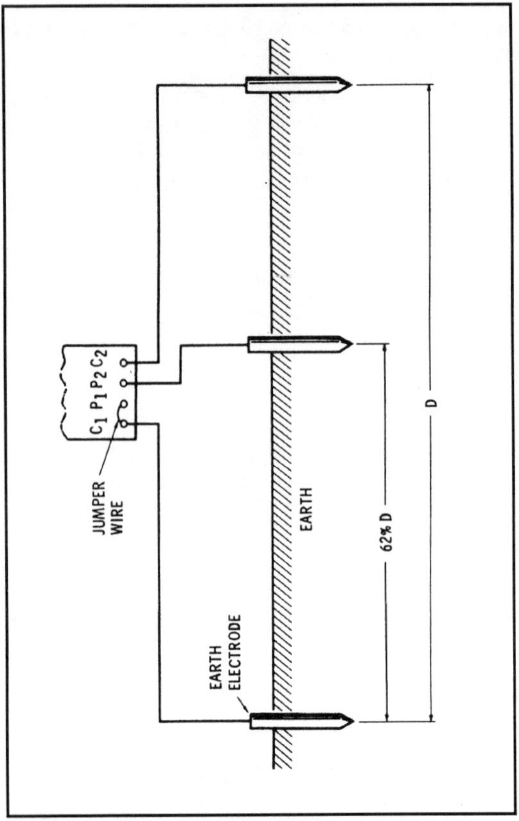

Figure 5-8: Fall-of-potential or three-terminal earth-resistance test.

and the rod connected to terminal C_2. Resistance readings are logged for each of the points. A curve of resistance versus distance (Figure 5-9) is then drawn. Correct earth resistance is read from the curve for the distance that is about 62 percent of the total distance from the earth electrode C_2; that is, if the total distance is D, then the 62 percent distance is $0.62\,D$. For example, if D is 120 feet, the distance value for earth resistance is 0.62×120 or 74 feet.

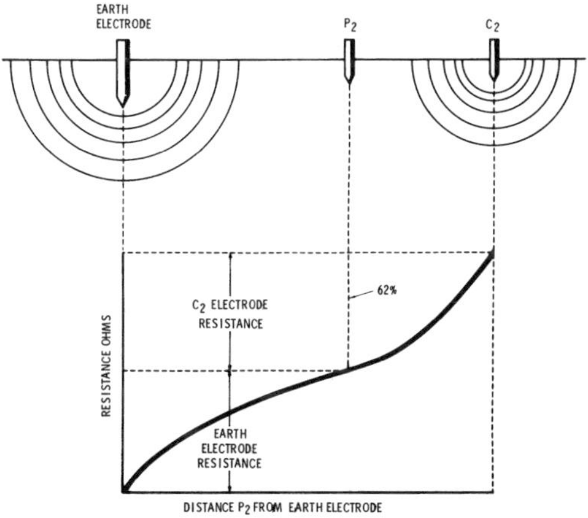

Figure 5-9: Resistance versus distance curve.

CHAPTER 6

Miscellaneous Testing Instruments

Frequency Meter

Frequency is the number of cycles completed each second by a given ac voltage; usually expressed in hertz; one hertz = 1 cycle per second.

The frequency meter is used in ac power-producing devices like generators to ensure that the correct frequency is being produced. Failure to produce the correct frequency will result in heat and component damage.

There are two common types of frequency meters. One operates with a set of reeds having natural vibration frequencies that respond in the range being tested. The reed with a natural frequency closest to that of the current being tested will vibrate most strongly when the meter operates. The frequency is read from a calibrated scale.

A moving-disk frequency meter works with two coils, one of which is a magnetizing coil whose current varies inversely with the frequency. A disk with a pointer mounted between the coils turns in the direction determined by the stronger coil. Solid-state frequency meters are also available.

Power-Factor Meter

The principle of operation of the power-factor meter is the principle of ratio meters which is also applied to synchroscopes, frequency meters, and some electrical thermometers.

Power factor is the ratio of the true power to the apparent power, and it depends on the phase difference between the current and the voltage. The power factor of a circuit can be obtained by taking simultaneous readings with an ammeter, a voltmeter, and a wattmeter and then dividing the wattmeter reading by the product of the voltage and current (volt-amperes); that is,

$$\text{PF (single-phase circuit)} = \frac{\textit{True Power (wattmeter reading)}}{\textit{Apparent Power (volt–amperes)}}$$

Much time, however, can be saved by using a single instrument — the power-factor meter. This instrument directly indicates the value of the power factor. For example, the single-phase power-factor meter in Figure 6-1 is so constructed that the rotating field is produced by the line voltage, and at a power factor of 100 percent, or unity, there is no torque on the moving coil and the pointer rests at the center of the scale, or 100. Depending on whether the current is leading or lagging when the power factor is less than unity, the pointer swings to the

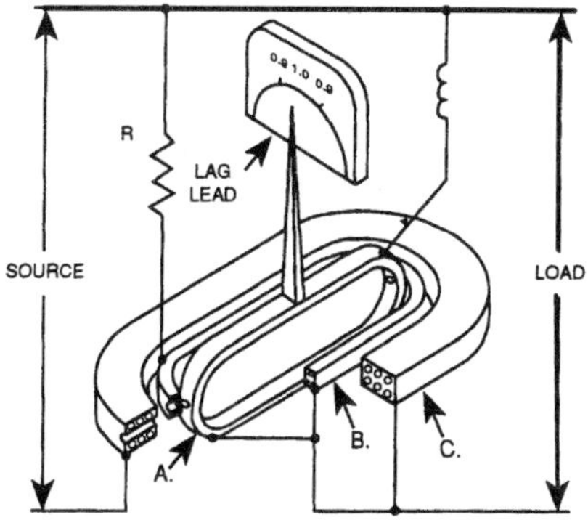

Figure 6-1: Power-factor meter.

left or to the right of the scale, indicating directly the value of the power factor.

Three-phase power-factor meters are also available and are quite common on industrial switchboard installations. Since most industrial establishments are charged a penalty if the power factor falls below 90 percent, industries try to maintain a high power factor at all times. A power-factor meter will continuously indicate the power factor so that a correction may be made at once, if necessary.

Power companies supplying power to industrial customers usually install two meters — one to measure active kilowatt hours and the other to measure reactive kilovar-hours; both are provided with 15-minute attachments. These demand attachments register the demand in kilowatts and kilovolt-amperes reactive for the two respective meters if sustained for a period of more than 15 minutes at any time. If the power-factor meter on the industrial switchboard shows that the power factor is below the 90 percent mark, an attempt can be made to correct it at once to avoid the penalty. Two other reasons for trying to maintain a power factor near unity are:

- The reduction of reactive current provides more capacity for useful current on the mains, feeds, and sub-feeders.
- A high power factor provides better voltage regulation and stability.

Synchroscopes

When two alternators are about to be connected in parallel, the voltages of the two must be approximately the same; their voltages must be exactly in phase; and their frequencies must be approximately the same. If these differences are too great, the alternators are likely to pull entirely out of phase, thus causing a complete shutdown. Therefore, some means of showing the difference in phase and frequency between the voltages of the two alternators is necessary.

Figure 6-2 on the next page shows a schematic drawing of the connections for a synchroscope in a high-voltage generating system. The synchroscope shown provides the following functions:

- It indicates whether the generator is running too slow or too fast.
- It indicates the amount by which the generator is slow or fast (the difference in frequency) at any instant.
- It indicates the exact time of coincidence in phase relationship between all generators connected to the system.

When the synchroscope shows that both alternators are synchronized, the switches may be closed to allow both alternators to work together in parallel. The ideal indication, on most synchroscopes, is when the pointer either stops or moves very slowly at mid-scale.

Tachometers

A tachometer is a device used to indicate or record the speed of a machine in revolutions per minute (rpm). Several designs are available which include the following:

- Centrifugal
- Eddy current
- Surface speed

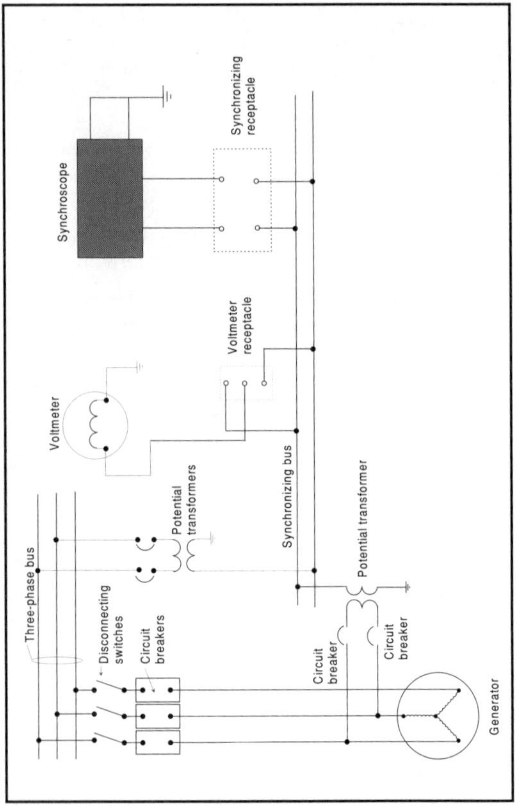

Figure 6-2: Synchronizing high-voltage generating system.

- Vibrating reed
- High-intensity stroboscope or photo tachometer

Some of the tachometers mentioned are connected to a motor, generator, or other machine by means of belts or gears while others are hand held. All have various scales from 0 rpm to over 50,000 rpm. Most of the better hand-held models have two buttons; a stop button to hold readings and a second button to release the pointer to 0. The ball-bearing spindle on these devices is placed against a rotating object on the machine such as the motor-drive shaft, and the speed of the shaft is read directly on the dial of the tachometer.

Totally enclosed rotating equipment may be checked for speed by using a vibrating-reed tachometer. These tachometers operate on the well-known and time-tested principle of resonance. The instrument is simply held against the motor, turbine, pump, compressor, or other rotating equipment and the speed is shown by the vibration of a steel reed which is tuned to a certain standard speed.

A photo tachometer uses a light that is aimed at the rotating shaft on which there is a contrasting color such as a mark, a chalk line, or a light-reflective strip or tape. The rotational speed in rpm is conveniently read directly from the indicating scale of the instrument. This tachometer design is especially useful on relatively inaccessible rotational equipment such as motors, fans, grinding wheel, and other similar machines. This is a definite advantage in many tests where it is difficult, if not impossible, to make contact with the rotational unit. Most photo tachometers can be recalibrated in the field, if necessary, and are powered by long-life mercury batteries.

An electrical tachometer consists of a small generator that is belted or geared to a unit whose speed is to be measured. The voltage produced in the generator varies directly with the speed of the rotating part of the generator. Since this speed is directly proportional to the speed of the machine under test, the amount of the generated voltage is a measure of the speed. The generator is electrically connected to an indicating or recording instrument which is calibrated to indicate units of speed such as rpm, fps, fpm, etc.

MISCELLANEOUS TESTING

There are a number of instruments that are used for testing electrically-operated equipment. Most will not be used frequently by the average electrician, but they are worth mentioning to acquaint you with their characteristics.

Footcandle Meter

There are a number of large and elaborate devices used in laboratories for making exact test and measurement of lamps and lighting fixtures. But for practical use in the field, a portable light or footcandle meter is quite satisfactory.

A typical footcandle meter consists of light-barrier layer cells and a meter enclosed in a suitable covering that is capable of reading light intensities from 1 to 500 footcandles or more.

To use the footcandle meter described, first remove the cover. Hold the meter in a position so the cell is facing toward the light source and at the level of the work plane where the illumination is required. The shadow of your body should not be allowed to fall on the cell during tests. A number of such tests at various points in a room or area will give the average illumination level in footcandles. Readings are taken directly from the meter scale.

Electrical Thermometers

For the measurement of temperatures, three basic electrical thermometer methods are used:

The resistance method which makes use of the fact that the resistance of a metal varies in direct proportion to temperature. This method is normally used for temperatures up to approximately 1500°F.

The thermocouple method which is based on the principle that a difference in temperature in different metals generates a voltage and is used for measuring temperatures up to about 3000°F.

The radiation-pyrometer and optical-pyrometer are generally used for temperatures above 3000°F. They combine the principle of the thermocouple with the effect of radiation of heat and light.

Phase-Sequence Indicator

A common phase-sequence indicator is designed for use in conjunction with any multimeter that can measure ac voltage. Most can be used on circuits with line voltages up to 550 volts ac, provided the instrument used with the indicator has a rating this high.

To use the phase-sequence indicator, set the multimeter to the proper voltage range. This can be determined (if it is not known) by measuring the line voltage before connecting the phase-sequence indicator. Next, connect the two black leads of the indicator to the voltage test leads of the meter. Connect the red, yellow, and black adapter leads to the circuit in any order and check the meter for voltage reading. If the meter reading is higher than the original circuit voltage measured, then the phase sequence is black-yellow-red. If the meter reading is lower than the original circuit voltage measured, then the phase sequence is red-yellow-black. If the reading is the same as the first reading, then one phase is open.

Infrared Sensing Device

An infrared sensing device is an optical device that measures the infrared heat emitted from an object.

Cable-Length Meter

A cable-length meter measures the length and condition of a cable by sending a signal down the cable and then reading the signal that is reflected back.

Digital Meters

The accuracy of digital meters, in many cases, far exceeds that of analog or dial-type meters. Better than 0.5% accuracy is fairly standard for most digital meters. This improved accuracy is due to the carefully calibrated electronic circuitry contained in digital meters. Because of the readout on a digital meter, the precision of the reading accuracy by the operator is exact — not a close estimate.

Digital meters also offer an advantage in poor lighting conditions when the face of an analog meter might be hard to see. LEDs (light-emitting diodes) displays can easily be seen in low-light areas. They do, however, consume more power and, therefore, the batteries may not last as long.

Digital meters are also smaller and are often less expensive to manufacture than the bulky analog meters of the past. Consequently, digital meters are quickly replacing the dial-type meters and you will probably see more and more digital type meters appear on job sites than any other types.

Summary

The use of test equipment is an important part of every technician's job. Selecting the proper instrument to be used in a specific application will help the worker to fully perform his or her task. The electrician will use the information provided by these instruments to help evaluate the work being accomplished.

Part 2

TROUBLESHOOTING AND MAINTENANCE OF ELECTRICAL SYSTEMS

CHAPTER 7

Troubleshooting Bearings

Alternating-current motors account for a high percentage of electrical repair work. And most of these failures can be traced to faulty bearings. Consequently, most industrial establishments place heavy emphasis on the proper handling, repair, and maintenance of various types of sleeve and ball bearings. As a result, plants are finding that electric motors last longer and perform better when a carefully planned motor lubrication schedule is followed.

If an ac motor failure occurs, the first step is to find out why. There are various causes of motor breakdowns, such as excessive load, binding or misalignment of motor drives, wet or dirty surroundings, and bearing failure. Bearing failure can occur in newer motors with high-quality bearings as frequently as in older motors equipped with less reliable bearings. A notable exception is motors equipped with sealed bearings which are much less prone to failure.

Considering this information, particularly the fact that sealed bearings are shielded from contamination and do not require lubrication, it stands to reason that contamination of bearings is one of the major causes of bearing failure. Consequently, most plant maintenance departments are updating the lubrication methods and bearing maintenance techniques, emphasizing cleanliness in procedures for all types of motor bearings.

Types of Bearings

There are many types of bearings, but ball bearings seem to be the most common. This type of bearing is found on various-size motors and their construction may be:

- Open
- Single shielded
- Double shielded
- Sealed
- Double row and other special types

Open bearings, as the name implies, are open construction and must be installed in a sealed housing. These bearings are less apt to cause churning of grease, and are therefore used mostly on large motors.

The single-shield bearing has a shield on one side to preclude grease from the motor windings. Double-shielded bearings have a shield on both sides of the bearing. This type of bearing is less susceptible to contamination and, because of its design, reduces the possibility of over-greasing. Sealed bearings have, on each side of the bearing, double shields which form an excellent seal. This bearing requires no maintenance, affords protection from contamination at all times, and does not require regreasing. It is normally used on small or medium-size motors.

The largest motors usually are furnished with oil-ring sleeve bearings. And some of the fractional-horsepower motors are equipped with plain sleeve bearings.

Each bearing type has characteristics which make it the best choice for a certain application. Replacement should be made with the same type bearings. The following list of functions provide a basic understanding of bearing application, a guide to analysis of bearing troubles due to misapplication, and emphasize the importance of proper replacement.

Figure 7-1 shows several types of bearings used in electric motors. The following is a brief description of each:

Self-aligning ball bearings: The self-aligning ball bearing, with two rows of balls rolling on the spherical surface of the outer ring, compensates for angular misalignment resulting from errors in mounting, shaft deflection, and distortion of the foundation. It is impossible for this bearing to exert any bending influence on the shaft — a most important consideration in applications requiring extreme accuracy, at high speeds. Self-aligning ball bearings are used for radial loads and moderate thrust loads in either direction.

Single-row, deep-groove ball bearings: The single-row, deep-groove ball bearing will sustain, in addition

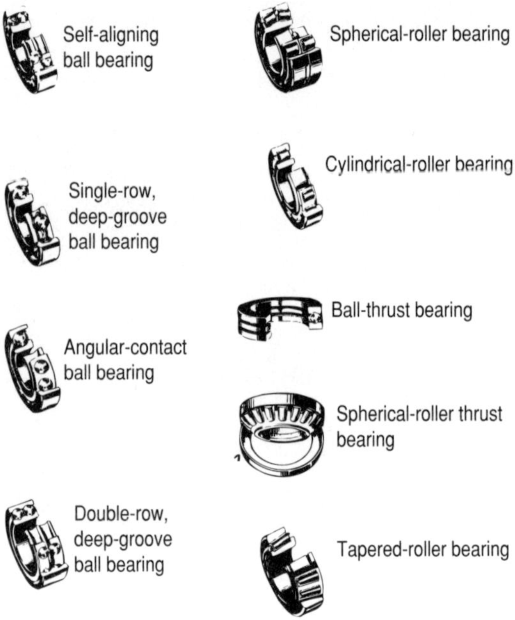

Figure 7-1: Various bearing types.

to radial load, a substantial thrust load in either direction, even at very high speeds. This advantage results from the intimate contact existing between the balls and the deep, continuous groove in each ring. When using this type of bearing, careful alignment between the shaft and housing is essential. This bearing is also available with seals and shields, which serves to exclude dirt and retain lubricant.

Angular-contact ball bearings: The angular-contact ball bearing supports a heavy thrust load in one direction, sometimes combined with a moderate radial load. A steep contact angle, assuring the highest thrust capacity and axial rigidity, is obtained by a high thrust-supporting shoulder on the inner ring and a similar high shoulder on the opposite side of the outer ring. These bearings can be mounted singly or, when the sides are flush ground, in tandem for constant thrust in one direction; mounted in pairs, also when sides are flush ground, for a combined load, either face-to-face or back-to-back.

Double-row, deep-groove ball bearings: The double-row, deep-groove ball bearing embodies the same principle of design as the single-row bearing. However, this bearing has a lower axial displacement than occurs in the single-row design, substantial thrust capacity in either direction, and high radial capacity due to the two rows of balls.

Spherical-roller bearings: The spherical-roller bearing, due to the number, size, and shape of the rollers, and the accuracy with which they are guided, has maximum capacity. Since the bearing is inherently self-aligning, angular misalignment between the shaft and housing has no detrimental effect, and the full capacity is always available for useful work. The design and proportion are such that, in addition to radial load, thrust loads may be carried in either direction.

Cylindrical-roller bearings: This type of bearing has high radial capacity and provides accurate guiding of the rollers, resulting in a close approach to true rolling. Consequent low friction permits operation at high speed. Those types which have flanges on one ring only, allow a limited free axial movement of the shaft in relation to the housing. They are easy to dismount even when both rings are mounted with a tight fit. The double-row type is particularly suitable for machine-tool spindles.

Ball-thrust bearings: The ball-thrust bearing is designed for thrust load in one direction only. The load line through the balls are parallel to the axis of the shaft, resulting in high-thrust capacity and minimum-axial deflection. Flat seats are preferred for heavy loads or for close axial positioning of the shaft.

Spherical-roller thrust bearings: The spherical-roller thrust bearing is designed to carry heavy thrust loads, or combined loads which are predominantly thrust. This bearing has a single row of rollers which roll on a spherical outer race with full self-alignment. The cage, centered by an inner ring sleeve, is constructed so that lubricant is pumped directly against the inner ring's unusually high guide flange. This bearing operates best with relatively heavy-oil lubrication.

Tapered-roller bearings: Since the axes of its rollers and raceways form an angle with the shaft axis, the tapered-roller bearing is especially suitable for carrying radial and axial loads acting simultaneously. A bearing of this type usually must be adjusted toward another bearing capable of carrying thrust loads in the opposite direction. Tapered-roller bearings are separable—their cones (inner rings) with rollers and their cups (outer rings) are mounted separately.

The do's and don'ts for ball-bearing assembly, maintenance, inspection, and lubrication are shown in Figure 7-2. Refer to this list often when working with electric motors.

Frequency Of Lubrication

Frequency of motor lubrication depends not only on the type of bearing but also on the motor application.

Small and medium-size motors equipped with ball bearings (except sealed bearings) are greased every three to six years if the motor duty is normal. On severe applications (high temperature, wet or dirty locations, or corrosive atmospheres), lubrication may be required more often. In severe applications, past experience and

DO's

DO work with clean tools, in clean surroundings.

DO remove all outside dirt from housing before exposing bearing.

DO treat a used bearing as carefully as a new one.

DO use clean solvents and flushing oils.

DO lay bearings out on clean paper or cloth.

DO protect disassembled bearings from dirt and moisture.

DO use clean, lint-free rags to wipe bearings.

DO keep bearings wrapped in oil-proof paper when not in use.

DO clean outside of housing before replacing bearings.

DO keep bearing lubricants clean when applying and cover containers when not in use.

DO be sure shaft size is within specified tolerances recommended for the bearing.

DO store bearings in original unopened cartons in a dry place.

DO use a clean, short-bristle brush with firmly embedded bristles to remove dirt, scale, or chips.

Figure 7-2: Do's and don'ts for ball-bearing assembly, maintenance, and lubrication.

DO's (Cont.)

DO be certain that, when installed, the bearing is square with and held firmly against the shaft shoulder.

DO follow lubricating instructions supplied with the machinery. Use only grease where grease is specified; use only oil where oil is specified. Be sure to use the exact kind of lubricant called for.

DO handle grease with clean paddles or grease runs. Store grease in clean containers. Keep grease containers covered.

DON'Ts

DON'T work under the handicap of poor tools, rough bench, or dirty surroundings.

DON'T use dirty, brittle, or chipped tools.

DON'T handle bearings with dirty, moist hands.

DON'T spin uncleaned bearings.

DON'T spin any bearings with compressed air.

DON'T use same container for cleaning and final rinse of bearings.

DON'T scratch or nick bearing surfaces.

Figure 7-2: Do's and don'ts for ball-bearing assembly, maintenance, and lubrication. ***(Cont.)***

DON'TS (*Cont.*)

DON'T remove grease or oil from new bearings.

DON'T use incorrect kind or amount of lubricant.

DON'T use a bearing as a gauge to check either the housing bore or the shaft fit.

DON'T install a bearing on a shaft that shows excessive wear.

DON'T open carton until bearing is ready for installation.

DON'T judge the condition of a bearing until after it has been cleaned.

DON'T pound directly on a bearing or ring, when installing, as this may cause damage to shaft and bearing.

DON'T overfill when lubricating. Excess grease and oil will ooze out of the overfilled housings past seals and closures, collect dirt, and cause trouble. Too much lubricant will also cause overheating, particularly where bearings operate at high speeds.

DON'T permit any machine to stand inoperative for months without turning it over periodically. This prevents moisture which may condense in a standing bearing from causing corrosion.

Figure 7-2: Do's and don'ts for ball-bearing assembly, maintenance, and lubrication. ***(Cont.)***

condition of the grease are the best guides as to frequency of lubrication.

Lubrication in sleeve bearings should be changed at least once a year. When the motor duty is severe or the oil appears dirty, it should be changed sooner.

Lubrication Procedure

For effective motor lubrication, cleanliness and use of the proper lubricant are of paramount importance.

When greasing a ball-bearing motor, the bearing housing, grease gun, and fittings are wiped clean. Great care must be taken to keep dirt out of the bearing when greasing. Next, the relief plug is removed from the bottom of the bearing housing. This is done to prevent excessive pressure from building up inside the bearing housing during greasing. Grease is then added, with the motor running if possible, until it begins to flow from the relief hole. Allow the motor to run from 5 to 10 minutes to expel excess grease. Then the relief plug is replaced and the bearing housing is cleaned.

It is important to avoid over-greasing. When too much grease is forced into a bearing, a churning of the lubricant occurs, resulting in high temperature and eventual bearing failure.

On motors that do not have a relief hole, grease should be applied sparingly. If possible, disassemble the motor and repack the bearing housing with the proper amount of grease. During this procedure, always maintain strict cleanliness.

The importance of adherence to these procedures cannot be overemphasized. Contamination and overgreasing of bearings are the major causes of bearing failure.

For sleeve bearings, use only the recommended oil for particular service conditions. Observing careful cleanliness, old oil is removed and new oil is added until the oil level reaches the "full" line on the oil sight gauge. This is done only when the motor is not running.

Testing Bearings

Two of the most effective tests are what might be called the "feel" test and the "sound" test. When performing the "feel" test, if, while the motor is running, the bearing housing feels overly hot to the touch, it is probably malfunctioning.

Note: Some bearings may operate safely up to about 85°C.

During the "sound" test, listen for foreign noises coming from the motor. Also, one end of a steel rod (about 3 feet long and ½ inch in diameter) may be placed on the bearing housing while the other end is held against the ear. The rod then acts as an amplifier, transmitting unusual sounds such as thumping or grinding, which would indicate a failing bearing. Special listening devices, such as the transistorized stethoscope, can also be used for the purpose.

Additional checks are usually in the form of checking the air gap on sleeve-bearing motors periodically. These tests, performed with a feeler gauge, indicate when a bearing begins to wear. Four measurements should be taken about 90 degrees apart around the rotor periphery. These measurements are recorded and compared with earlier readings, providing a check on the condition of bearings.

Motors should also be checked for end play. Ball-bearing motors should have about $\frac{1}{32}$ inch to $\frac{1}{16}$ inch end play. Sleeve-bearing motors may have up to $\frac{1}{2}$ inch end play.

On large sleeve bearings the oil level should be checked periodically, and the oil is visually inspected for contamination. If it is possible, the oil rings should be checked when the motor is operating.

Other inspections include checking for misaligned or bent shafts and for excessive belt pressure.

The troubleshooting chart in Figure 7-3 lists the most common problems with motor bearings.

Trouble	Cause	What to do
Hot Bearings — General	Bent or sprung shaft.	Straighten or replace shaft.
	Excessive belt pull.	Decrease belt tension.
	Pulley too far away.	Move pulley closer to bearing.
	Pulley diameter too small.	Use larger pulleys.
	Misalignment.	Correct by realignment of drive.

Figure 7-3: Troubleshooting chart for motor bearings.

Trouble	Cause	What to do
Hot Bearings — Sleeve	Oil grooving in bearing obstructed by dirt.	Remove bracket or pedestal with bearing and clean oil grooves and bearing housing; renew oil.
	Bent or damaged oil rings.	Repair or replace oil rings.
	Oil too heavy.	Use a recommended lighter oil.
	Oil too light.	Use a recommended heavier oil.

Figure 7-3: Troubleshooting chart for motor bearings. ***(Cont.)***

Trouble	Cause	What to do
Hot Bearings — Sleeve	Insufficient oil.	Fill reservoir to proper level in overflow plug with motor at rest.
	Too much end thrust.	Reduce thrust induced by driven machine or supply external means to carry thrust.
	Badly worn bearing.	Replace bearing.

Figure 7-3: Troubleshooting chart for motor bearings. ***(Cont.)***

Trouble	Cause	What to do
Hot Bearings — Ball	Insufficient grease.	Replace bearing.
	Deterioration of grease or lubricant contaminated.	Remove old grease, wash bearings thoroughly in kerosene and replace with new grease.
	Excess lubricant.	Reduce quantity of grease. Bearing should not be more than ½ filled.

Figure 7-3: Troubleshooting chart for motor bearings. ***(Cont.)***

Trouble	Cause	What to do
Hot Bearings — Ball	Heat from hot motor or external source.	Protect bearing by reducing motor temperature.
	Overloaded bearing.	Check alignment, side thrust and end thrust.
	Broken ball or rough races.	Replace bearing; first clean housing thoroughly.

Figure 7-3: Troubleshooting chart for motor bearings. *(Cont.)*

CHAPTER 8

Troubleshooting Contactors and Relays

The general classification of contactor covers a type of electromagnetic apparatus designed to handle relatively high currents.

A relay is an electromagnetic or solid-state device that is used in control circuits of magnetic motor starters, contactors, solenoids, timers, and other devices. Relays are generally used to amplify the contact capability or multiply the switching functions of a pilot device.

A tremendous variety of relays is used for lighting, motor, and HVAC control circuits, with the magnetic relay still being the most common, although solid-state control devices are rapidly finding their way into all types of applications.

The troubleshooting charts in this chapter give symptoms and the probable causes of problems that develop in magnetic contactors and relays. Knowing these symptoms and their cures should help spot and repair contactor and relay problems in a hurry.

Symptoms	Probable Cause	Action or Items to Check
Failure to pull in	Either no voltage or low voltage at coil terminals	Blown fuse, open line switch, break in wiring
		Line voltage below normal
		Overload relay open or set too low
		Tripping toggle (nonautomatic breakers) fouled
		Control level or start button in OFF position

Figure 8-1: Contactor and relay troubleshooting chart.

Symptoms	Probable Cause	Action or Items to Check
Failure to pull in	Either no voltage or low voltage at coil terminals	Pull-in circuit open, shorted or grounded
		Contacts in protective or controlling circuit open or one of their pigtail connections broken
	Operating coil open or grounded	Inspect and test coil
	Loose or disconnected coil lead wire	Inspect and correct

Figure 8-1: Contactor and relay troubleshooting chart. ***(Cont.)***

Symptoms	Probable Cause	Action or Items to Check
Failure to pull in	Excessive magnet gap, improper alignment	Inspect and correct
	Armature obstructed or gumming deposits between armature and pole face	Inspect and correct
	Binding caused by deformed or gummy hinge	Replace if bent; degrease if gummed up
	Excessive armature spring force	Lessen spring force

Figure 8-1: Contactor and relay troubleshooting chart. ***(Cont.)***

Symptoms	Probable Cause	Action or Items to Check
Failure to pull in	Normally closed contacts welded together	Replace contacts
Failure of equipment to start with contactor closed	One contact not closing	Replace set of contacts
	Contacts burned	Use emery cloth to lightly burnish or replace
	Contact pigtail connection broken	Replace

Figure 8-1: Contactor and relay troubleshooting chart. ***(Cont.)***

Symptoms	Probable Cause	Action or Items to Check
Failure to drop out	Operating coil is energized	Contacts in controlling or protective tripping circuits closed, shorted or shunted
		Tripping devices defective; such as, overload tripping toggles do not strike release, undervoltage relay plunger stuck or out of adjustment, defective stop button, defective time-delay escape mechanism (closed air vent)

Figure 8-1: Contactor and relay troubleshooting chart. ***(Cont.)***

Symptoms	Probable Cause	Action or Items to Check
Failure to drop out	Operating coil is energized	Current supplied over an unintentional path, due to grounds, defective insulation, pencil markings, moisture, lacquer chipped off relay's base
	Residual magnetism excessive, due to armature closed tightly against pole face	Adjust or replace

Figure 8-1: Contactor and relay troubleshooting chart. ***(Cont.)***

Symptoms	Probable Cause	Action or Items to Check
Failure to drop out	Armature obstructed or gumming deposits between armature and pole face	Clear and clean
	Binding caused by deformed or gummy hinge	Replace or degrease
	Contact pressure spring or armature spring too weak or improperly adjusted	Replace or adjust
	Improper mounting position (upside down)	Mount correctly

Figure 8-1: Contactor and relay troubleshooting chart. ***(Cont.)***

Symptoms	Probable Cause	Action or Items to Check
Failure to drop out	Normally opened contacts welded together	Replace contacts
Time delay relays operate too fast	Escapement mechanism faulty	Air escapes too freely due to holes in bellows or open air vent
		Worn dashpot plunger, dashpot oil too thin
	Non-magnetic shim in air gap too large or armature spring too strong	Adjust or replace
	Magnets out of adjustment	Adjust

Figure 8-1: Contator and relay troubleshooting chart. ***(Cont.)***

Symptoms	Probable Cause	Action or Items to Check
Contacts pitted or discolored	Contacts overheated from overload	Reduce load and replace contacts
	Contacts not fitted properly	Refit
	Barriers broken from rough usage or breakers closing with too much force	Replace barriers; check for high voltage
	Wiping action of contacts on closing is insufficient	Correct
	Excessive chatter or hum	Check shaded poles

Figure 8-1: Contactor and relay troubleshooting chart. ***(Cont.)***

Symptoms	Probable Cause	Action or Items to Check
Contacts pitted or discolored	Exposure to weather, dipping water, salt air or vibration	Use suitable NEMA enclosure
Excessive chatter or hum	Vibration of surrounding devices communicated to relay	Arrange differently
	Relay receiving contradictory signals	Correct
	Bouncing of controlling or protective contacts	Correct

Figure 8-1: Contactor and relay troubleshooting chart. ***(Cont.)***

Symptoms	Probable Cause	Action or Items to Check
Excessive chatter or hum	Line voltage fluctuating or insufficient	Consider using buck-and-boost transformers
	Excessive coil circuit resistance	Lessen resistance
	Armature spring or contact pressure spring too strong	Lessen
	Excessive coil voltage drop on closing	Correct
	Improper anti-freeze pin location	Locate in correct position

Figure 8-1: Contactor and relay troubleshooting chart. ***(Cont.)***

Symptoms	Probable Cause	Action or Items to Check
Excessive chatter or hum	Free movement of armature hindered due to deformed parts or dirt	Clean and replace appropriate parts
Excessive coil temperature	Excessive current or voltage	Reduce load or adjust taps on transformer
	Short circuit in coil	Replace coil
	Excessive eddy current and hysteresis	Correct
	High room temperature	Replace with properly designed relay

Figure 8-1: Contactor and relay troubleshooting chart. ***(Cont.)***

CHAPTER 9

Troubleshooting Dry-Type Transformers

Since transformers are an essential part of every electrical installation, electricians must know how to test and locate troubles that develop in transformers — especially in the smaller dry-type, power-supply or control transformers.

The many troubleshooting charts available are very useful when troubleshooting. To use troubleshooting charts, most list the complaint on the left side of the charts, then the possible cause, followed by the proper corrective action.

Think Before Acting: Study the problem thoroughly, then ask yourself these questions:

- What were the warning signs preceding the trouble?
- What previous repair and maintenance work has been done?
- Has similar trouble occurred before?
- If the circuit, component, or piece of equipment still operates, is it safe to continue operation before further testing?

The answers to these questions can usually be obtained by:

- Questioning the owner of the equipment.
- Taking time to think the problem through.
- Looking for additional symptoms.
- Consulting troubleshooting charts.
- Checking the simplest things first.
- Referring to repair and maintenance records.
- Checking with calibrated instruments.
- Double checking all conclusions before beginning any repair on the equipment or circuit components.

Note: Always check the easiest and obvious things first; following this simple rule will save time and trouble.

Double-Check Before Beginning

The source of many problems can be traced not to one part alone but to the relationship of one part with another. For instance, a tripped circuit breaker may be reset to restart a piece of equipment, but what caused the breaker to trip in the first place? It could have been caused by a vibrating "hot" conductor momentarily coming into contact with a ground, or a loose connection could eventually cause overheating, and any number of other causes.

Too often, electrically-operated equipment is completely disassembled in search of the cause of a certain complaint and all evidence is destroyed during disassembly operations. Check again to be certain an easy solution to the problem has not been overlooked.

Find and Correct Cause of Trouble

After an electrical failure has been corrected in any type of electrical circuit or piece of equipment, be sure to locate and correct the cause so the same failure will not be repeated. Further investigation may reveal other faulty components.

Also be aware that although troubleshooting charts and procedures greatly help in diagnosing malfunctions, they can never be complete. There are too many variations and solutions for a given problem.

To solve electrical problems consistently, you must first understand the basic parts of electrical circuits, how they function, and for what purpose. If you know that a particular part is not performing its job, then the cause of the malfunction must be within this part or series of parts.

TROUBLESHOOTING TRANSFORMERS

Open Circuit: Should one of the windings in a transformer develop a break or "open" condition, no current can flow and therefore, the transformer will not deliver any output. The symptoms of an open-circuited transformer is that the circuits which derive power from the transformer are de-energized or "dead." Use an ac voltmeter to check across the transformer output terminals as shown in Figure 9-1. A reading of zero volts indicates an open circuit.

Then take a voltage reading across the input terminals. If a voltage reading is present, then the conclusion is that one of the windings in the transformer is open.

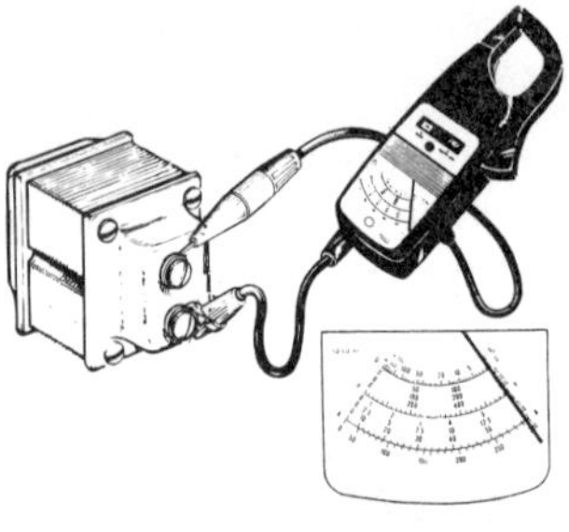

Figure 9-1: Checking for an open circuit in a transformer.

However, if no voltage reading is on the input terminals either, then the conclusion is that the open is elsewhere on the line side of the circuit; perhaps a disconnect switch is open.

WARNING!

Make absolutely certain that your testing instruments are designed for the job and are calibrated for the correct voltage. Never test the primary side of any transformer over 600 volts unless you are qualified, have the correct high-voltage testing instruments, and the test is made under the proper supervision.

However, if voltage is present on the line or primary side and none on the secondary or load side, open the switch to de-energize the circuit, and place a warning tag (tag-out and lock) on this switch so that it is not inadvertently closed again while someone is working on the circuit. Disconnect all of the transformer primary and secondary leads, check each winding in the transformer for continuity (a continuous circuit), as indicated by a resistance reading taken with an ohmmeter.

Continuity is indicated by a relatively low resistance reading on control transformers, while an open winding will be indicated by an infinite resistance reading on the ohmmeter. In most cases, such small transformers will have to be replaced, unless of course the break is accessible and can be repaired.

Ground Fault: Sometimes a few turns in the secondary winding of a transformer will acquire a partial short, which in turn will cause a voltage drop across the secondary. The symptom of this condition is usually overheating of the transformer caused by large circulating currents flowing in the shorted windings.

The easiest way to check this condition is with a voltmeter set at the proper voltage scale. Take a reading on the line or primary side of the transformer first to make certain normal voltage is present. Then take a reading on the secondary side. If the transformer has a partial short or ground fault, the voltage reading should be lower than normal.

Replace the faulty transformer with a new one and again take a reading on the secondary. If the voltage reading is now normal and the circuit operates satisfactorily, leave the replacement transformer in the circuit, and either discard or repair the original transformer.

A highly sensitive ohmmeter may also be used to test this condition when the system is de-energized and the leads are disconnected; a lower reading on the ohmmeter than normal indicates this condition. However, the reading will usually be so slight that the average ohmmeter is not sensitive enough to detect the difference. Therefore, the recommended way is to use the voltmeter test.

Complete Short: Occasionally a transformer winding will become completely shorted. In most cases, this will activate the overload-protective device and de-energize the circuit, but in other instances, the transformer may continue trying to operate with excessive overheating — due to the very large circulating current. This heat will often melt the wax or insulation inside the transformer, which is easily detected by the odor. Also, there will be no voltage output across the shorted winding and the circuit across the winding will be dead.

The short may be in the external secondary circuit or it may be in the transformer's winding. To determine its location, disconnect the external secondary circuit from the winding and take a reading with a voltmeter. If the voltage is normal with the external circuit disconnected, then the problem lies within the external circuit. However, if the voltage reading is still zero across the secondary leads, the transformer is shorted and will have to be replaced.

Grounded Windings: Insulation breakdown is quite common in older transformers — especially those that have been overloaded. At some point, the insulation breaks or deteriorates and the wire becomes exposed. The exposed wire often comes into contact with the transformer housing and grounds the winding.

If a winding develops a ground, and a point in the external circuit connected to this winding is also grounded, part of the winding will be shorted out. The symptoms will be overheating, which is usually detected by feel or smell, and a low voltage reading as indicated on a voltmeter scale. In most cases, transformers with this condition will have to be replaced.

A megohmmeter is the best test instrument to check for this condition. Disconnect the leads from both the primary and secondary windings. Tests can then be performed on either winding by connecting the megger negative test lead to an associated ground and the positive test lead to the winding to be measured.

Insulation resistance should then be measured between the windings themselves. This is accomplished by connecting one test lead to the primary and the second test lead to the secondary. All such tests should be recorded on a record card under proper identifying labels.

The troubleshooting chart in Figure 9-2 covers the most common dry-type transformer problems.

Malfunction	Probable Cause
Overheating	Continuous overload; wrong external connections; poor ventilation; high surrounding air temperatures

Figure 9-2: Troubleshooting chart for dry-type transformers.

Malfunction	Probable Cause
Overheating	High input voltage
	Clogged air ducts or inadequate ventilation
Reduced to zero voltage	Short turns; loose connections to transformer terminal board
Excess secondary voltage	Input voltage high; dirt accumulations on primary terminal board
High conductor loss	Overload; terminal boards not on identical tap position
Coil distortion	Coils short-circuited
Insulation failure	Continuous overloads; dirt accumulations on coils; mechanical damage in handling; lightning surge
	Very high core temperature due to high input voltage or low frequency
Breakers or fuses open	Short circuit; overload

Figure 9-2: Troubleshooting chart for dry-type transformers. ***(Cont.)***

Malfunction	Probable Cause
Excessive cable heating	Improperly bolted connection
High voltage to ground	Usually a static charge condition
Vibration and noise	Low frequency; high-input voltage; core clamps loosened in shipment or handling; loose hardware on enclosure; location
High exciting current	Low frequency; high input voltage; shorted turns (windings)
High core loss	Low frequency; high input voltage
Smoke	Insulation failure
Burned insulation	Lightning surge; switching or line disturbance; broken bushings

Figure 9-2: Troubleshooting chart for dry-type transformers. ***(Cont.)***

CHAPTER 10

Troubleshooting Fluorescent Fixtures

The fluorescent lamp (Figure 10-1) is an electrical discharge lighting source where a mercury arc generates ultraviolet energy which in turn activates the phosphor coatings on the inside of the tube to produce light. The lamp takes a variety of shapes, the dominant one being a smooth, long, tubular shape of various diameters and lengths. Due to their negative resistance, fluorescent lamps require a ballast to start and limit the current flow through them. In all cases, the lamps in a fluorescent fixture must match the ballast, and vice versa.

Preheat Lamps

Preheat lamps have bipin bases as shown in Figure 10-1 and are manufactured in sizes from 4 to 100 W and in tube lengths from 6 to 60 in long. Either manual or automatic starters are required to produce a separate cycle for cathode heating prior to operation. Some preheat lamps may be operated on *trigger start* circuits with *rapid start* starting characteristics. In either case, lamp operating characteristics are based upon their operation with ballasts.

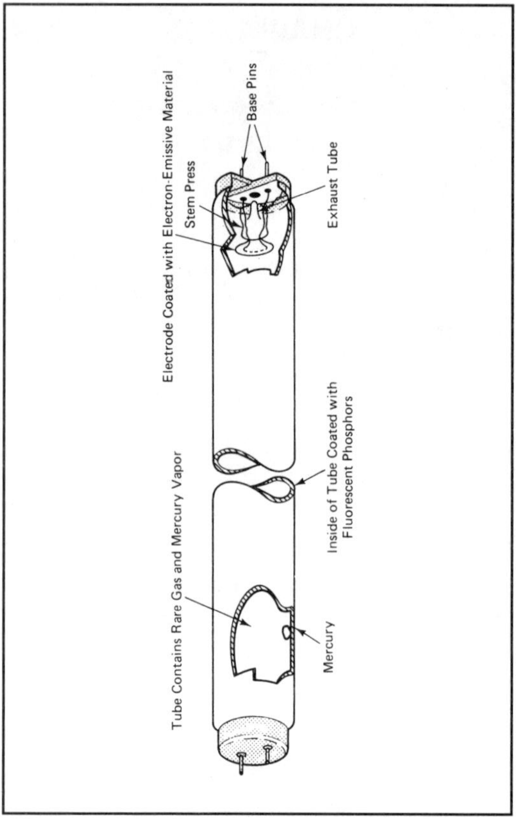

Figure 10-1: Basic components of a fluorescent lamp.

Rapid Start Lamps

Rapid start lamps require a preheat cycle, but unlike preheat lamps, do not require a separate starter. The types of lamps falling under this classification are as follows:

- *Rapid start bipin* — limited to 48 in, 40-W T-12 configuration. Operates at 430 mA.
- *High-output* — recessed double contact available in the range of 24 to 96 in and from 35 to 110 W; operates at current of 800 mA.
- *Power groove* — recessed double contact lamp bases available in this range of 48 to 96 in, and from 110 to 215 W in T-17 bulb size; operates at current of 1500 mA.
- *VHO (very high output) and SHO (super high output)* — manufacturers' brand names for recessed double contact lamp bases in the range of 48 to 96 in, from 110 to 215 W in T-12 bulb size, and operates at current of 1500 mA.

Every fluorescent lamp and starter should be tested before installation because lamps and starters are fragile and may possibly be damaged in shipment or handling. This will reveal any defectives and will prevent wasted time in installation, or in maintaining equipment in service. The troubleshooting charts to follow list faults, probable causes, and corrective action to take while troubleshooting fluorescent lighting fixtures and lamps.

Symptoms	Probable Cause	Action or Items to Check
Blinking on and off, along with shimmering effect during lighted period.	Normal end of lamp life, emission material on electrode depleted.	Replace lamp.
Blinking of relatively new lamps	Incorrect or defective starter. Loose circuit contact.	Replace starter. Seat lamp securely; indicator bumps should be directly over socket slot. Check if lamp holders are rigidly mounted and properly spaced; tighten all connections.

Figure 10-2: Troubleshooting chart for fluorescent lamp equipment.

Symptoms	Probable Cause	Action or Items to Check
Blinking with two-lamp ballasts. One lamp starts and one end of the other may blink on and off without starting; eventually both lamps may start.	Cold drafts hitting lamp. Starter leads improperly wired.	Enclose or protect lamp. Check wiring diagram on ballast and reconnect leads correctly or interchange lamp holders at one end of fixture.
Ends of lamp remain lighted.	Starter contacts are stuck together.	Replace starter.

Figure 10-2: Troubleshooting chart for fluorescent lamp equipment. *(Cont.)*

Symptoms	Probable Cause	Action or Items to Check
No starting effort, or slow starting.	Open circuit in electrodes or air leak in lamp.	Check lamps in another fixture; if not fluorescent end glow exists, replace them.
Slow starting of preheat type lamps.	Starter sluggish or at end of life.	Replace starter.
	Possible open circuit.	Check wiring.
	No starting compensator in leading circuit of two-lamp ballast.	Install one in series with starter.
	Low-ballast rating.	Replace with ballast of proper rating.
	Low circuit voltage.	Check voltage and correct if possible.

Figure 10-2: Troubleshooting chart for fluorescent lamp equipment. ***(Cont.)***

Symptoms	Probable Cause	Action or Items to Check
Slow starting of preheat type lamps.	Remote possibility of open-circuited ballasts.	Check ballast.
Slow starting of rapid start type lamps during conditions of high humidity.	Dust or dirt on lamps overcomes effect of silicone coating.	Wash lamps in water containing mild detergent. Rinse lamps with clean water to prevent film from depositing on bulb.
Very slow life accompanied by severe end blackening on rapid start type lamps.	Electrode not being heated.	Check for poor contact between lamp pins and socket. Check for open circuit — loose socket contacts or break in line.

Figure 10-2: Troubleshooting chart for fluorescent lamp equipment. *(Cont.)*

Symptoms	Probable Cause	Action or Items to Check
Very slow life accompanied by severe end blackening on rapid start type lamps.	Loose circuit contact causing on-off blink.	Be sure the lamp holders are rigidly mounted and that the lamp is securely seated; check the circuit wiring.
	Too low or too high voltage.	Check the line voltage and be sure it is within the range on the ballast nameplate.
	Low ambient temperature.	Use special auxiliaries for temperature below 50°F.
	Ballast prematurely starting preheat lamps.	Replace starter with a thermal type.

Figure 10-2: Troubleshooting chart for fluorescent lamp equipment. ***(Cont.)***

Symptoms	Probable Cause	Action or Items to Check
Very short life accompanied by severe end blackening on rapid start type lamps.	No starter compensator in leading circuit of two-lamp ballast.	Install one in series with starter.
	In series instant-start ballasts, one lamp burned out and the other burning dimly. If the burned out lamp is not replaced, the dim lamp will burn out shortly.	Replace the burned out lamp immediately.
	Series instant start ballasts, both lamps out, only one lamp may be defective.	Check both lamps to determine which is defective, reinstall good lamp.

Figure 10-2: Troubleshooting chart for fluorescent lamp equipment. ***(Cont.)***

Symptoms	Probable Cause	Action or Items to Check
Very short lamp life accompanied by severe end blackening on lamps.	Ballast improperly designed or not within specifications or wrong ballast being used.	Use CBM certified ballast of correct rating for lamp size.
	Improper ballast equipment of dc sources.	Check ballast equipment.
Dense blackening at one or both ends of tube, extending two to three inches from base.	Normal end of lamp life.	Replace lamp.

Figure 10-2: Troubleshooting chart for fluorescent lamp equipment. ***(Cont.)***

Symptoms	Probable Cause	Action or Items to Check
Dense blackening at one or both ends of tube, extending two to three inches from base.	With rapid-start lamps, accompanied by short life — poor contact between lamp pins and socket.	Check for proper socket spacing or poor socket construction not providing proper wiping of pins when lamp is installed.
Blackening generally within one inch of ends.	Mercury deposit.	Should evaporate as lamp is operated.
Blackening early in life.	Starter defective, causing on-off blink or prolonged flashing.	Replace starter.

Figure 10-2: Troubleshooting chart for fluorescent lamp equipment. ***(Cont.)***

Symptoms	Probable Cause	Action or Items to Check
Blackening early in life.	Ends of lamp remain lighted because of starter failure.	Replace starter.
	Too low or too high voltage.	Check the line voltage to be certain it is within the range shown on the ballast nameplate.
	Loose circuit contact causing on-off blink.	Be sure the lamp holders are rigidly mounted and the lamp is securely seated; check circuit wiring.
	No starting compensator in leading circuit.	Install one in series with starter.

Figure 10-2: Troubleshooting chart for fluorescent lamp equipment. ***(Cont.)***

Symptoms	Probable Cause	Action or Items to Check
Blackening early in life.	Ballast improperly designed or not within specifications or wrong ballast being used.	Use CBM certified ballast of correct rating for lamp size.
Dense spot — black about ½ inch wide, extending about half way around the tube, centering about one inch from the base.	Normal sign of age in service. If early in life, indicates excessive lamp starting or high operating current.	Check for off rating or ballast or unusually high-current voltage. Ballast may be improperly designed.

Figure 10-2: Troubleshooting chart for fluorescent lamp equipment. ***(Cont.)***

Symptoms	Probable Cause	Action or Items to Check
Rings — brownish ring at one end or both, about two inches from base.	This may develop on some lamps during operation.	Will not affect the lamp performance.
Dark streaks lengthwise of tube.	Globules of mercury condensed on lower part of tube.	The lower half of the tube is cooler than the upper half; by rotating the tube 180° these mercury globules should evaporate due to the increased warmth.

Figure 10-2: Troubleshooting chart for fluorescent lamp equipment. ***(Cont.)***

Symptoms	Probable Cause	Action or Items to Check
Pronounced swirling, spiraling or fluttering of arc stream.	May occur in new lamps.	Should season out in normal operation.
	Starter not performing properly to preheat electrodes.	Replace starter.
	No starting compensator in leading circuit of two-lamp ballast.	Install one in series with starter.
	Ballast improperly designed or not within specifications or wrong ballast used.	Use CBM certified ballasts of correct rating for lamp size.

Figure 10-2: Troubleshooting chart for fluorescent lamp equipment. *(Cont.)*

Symptoms	Probable Cause	Action or Items to Check
Pronounced swirling, spiraling or fluttering of arc stream.	High-voltage starting.	Check voltage and correct, if possible. If condition persists with good operating condition, replace lamp.
Radio interference.	Lamp-radiation broadcasts through radio receiver.	Small condenser in starter or ballast may be defective; replace.
	Line radiation and line feedback.	Apply radio-interference filter at lamp or fixture.

Figure 10-2: Troubleshooting chart for fluorescent lamp equipment. ***(Cont.)***

Symptoms	Probable Cause	Action or Items to Check
Noise: humming sound that may be steady or come and go.	Internal variation in ballast.	Tighten fixture louvers, glass, side panels, etc. If ballast continues to be noisy, replace it.
	Overheated ballasts.	Prolonged blinking of lamp tends to heat ballast; replace ballast.
Dark section of tube — ¼ to ½ of tube gives no light.	DC operation without using reversing switches.	Install reversing switches.
Decreased light output.	Cold drafts hitting tube.	Enclose or protect lamp. Better ventilation of fixture.

Figure 10-2: Troubleshooting chart for fluorescent lamp equipment. ***(Cont.)***

Symptoms	Probable Cause	Action or Items to Check
Decreased light output.	Low-temperature operation.	Enclose the lamp.
	Low circuit voltage.	Check voltage and correct if possible.
	Dust or dirt on lamp, fixture, walls, or ceilings.	Clean.
Lamps operate at unequal brilliancy.	Low circuit voltage on two-lamp lead-lag ballasts.	Check voltage and correct if possible. Possibly defective ballast.
Different color appearance in difference areas of installation.	Holders spaced wrong.	Correct fixture dimensions.

Figure 10-2: Troubleshooting chart for fluorescent lamp equipment. *(Cont.)*

CHAPTER 11

Troubleshooting Incandescent Lamps and Fixtures

Incandescent lamps are made in thousands of different types and colors from a fraction of a watt to over 10 kW each, and for practically any conceivable lighting application.

Regardless of the type or size, all incandescent filament lamps consist of a sealed glass envelope containing a filament. Light is produced when the filament is heated to incandescence (white light) by its resistance to a flow of electric current. Figure 11-1 shows the basic components of an incandescent lamps.

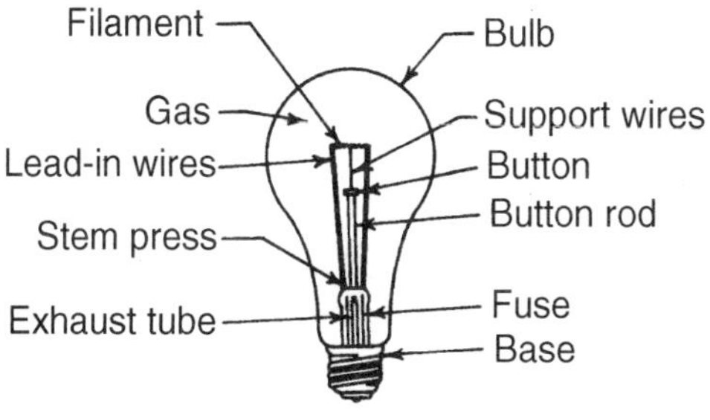

Figure 11-1: Basic components of an incandescent lamp.

The quartz-iodine tungsten-filament lamp is basically an incandescent lamp, since light is produced from the incandescence of its coiled tungsten filament. However, the quartz lamp envelope is filled with an iodine vapor, which prevents the evaporation of the tungsten filament. This evaporation is what normally occurs in conventional incandescent lamps; then the bulb begins to blacken, light output deteriorates, and eventually the filament burns out.

While the quartz-iodine lamp has approximately the same efficiency as an equivalent conventional incandescent lamp, it has the advantages of double the normal life, low lumen depreciation, and a smaller bulb for a given wattage. Figure 11-2 illustrates the basic components of a quartz-iodine lamp.

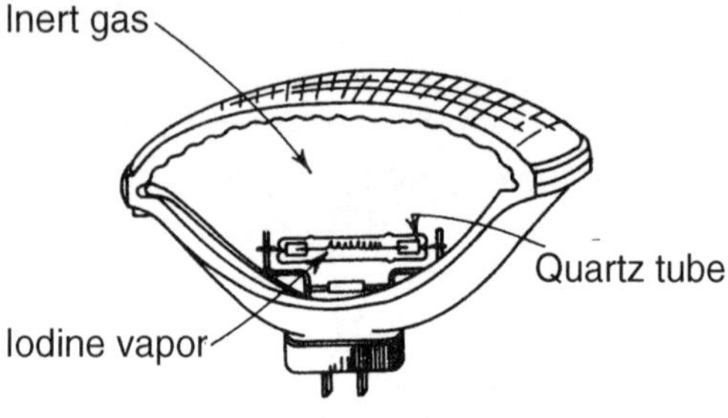

Figure 11-2: Basic components of a quartz-iodine tungsten-filament lamp.

The troubleshooting charts to follow (Figure 11-3) cover the most commonly encountered problems with incandescent lamps and lighting fixtures.

Symptoms	Probable Cause	Action or Items to Check
Lamp not burning, but apparently okay.	Lamp loose.	Tighten in socket.
	Loose or broken connections.	Secure terminals.
		Repair wiring.
Lamp burns dim.	Low voltage.	Match lamp rating to line voltage.
		Increase line voltage.
Short lamp life.	High voltage.	Match lamp rating to line voltage.
	Bulb cracked due to mechanical shock.	Replace lamp.
		Make sure that water does not drip on bulb.

Figure 11-3: Troubleshooting chart for incandescent lamp equipment.

Symptoms	Probable Cause	Action or Items to Check
Short lamp life.	Incorrect lamp.	Replace with lamp of size for which luminaire is rated.
	Excessive vibration.	Use shock absorbing device.
Lamp breakage.	Water contacts lamp bulb.	Use enclosed, vaportight luminaire if water spray is present.
		Seal joint where conduit stem enters luminaire.
	Bulb touches luminaire.	Use correct size of lamp.
		Straighten socket.

Figure 11-3: Troubleshooting chart for incandescent lamp equipment. ***(Cont.)***

CHAPTER 12

Troubleshooting HID Lamps and Fixtures

HID (high-intensity discharge) lamp is a generic term which describes a wide variety of lighting sources. All types of HID lamps consist of gaseous discharge arc tubes which, in the versions designed for lighting, operate at pressures and current densities sufficient to generate desired quantities of radiation within their arcs alone.

Mercury vapor lamps contain arc tubes which are formed of fused quartz. This construction has resulted in great improvements in lamp life and maintenance of output through life. These arcs radiate ultraviolet energy as well as light, but the glass used in outer bulbs is generally of a heat-resisting type that absorbs most of the ultraviolet. Some mercury lamps have outer bulbs that are internally coated with fluorescent materials which, when activated by the ultraviolet, emit visible energy at wavelengths that modify the color of light from the arc. The General Electric deluxe white mercury lamps, for example, have color characteristics that are well suited to many commercial lighting applications that could not, until recently, have been considered for mercury. General lighting mercury lamps are available in wattages for 50 to 3000 watts.

Multivapor lamps generate light with more than half again the efficiency of the mercury arc, and with better color quality.

The outer bulbs of HID lamps are designed to provide, as nearly as possible, optimum internal environments for arc-tube performance. For example, the rounded shapes labeled E and BT in Figure 12-1 were

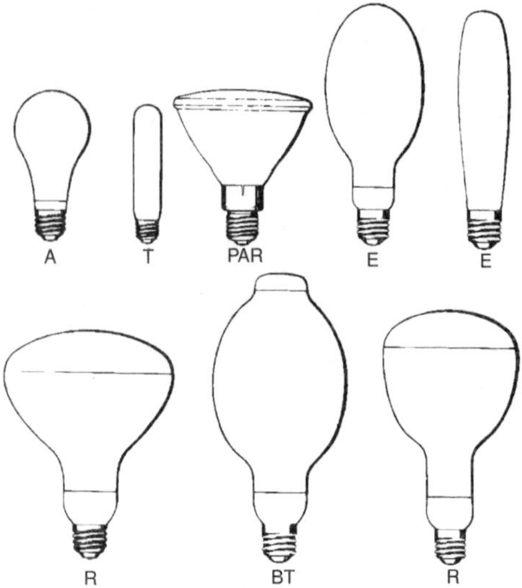

Figure 12-1: The rounded shaped bulbs labeled E and BT were devised to maintain uniform temperatures of the bulb walls for better performance of phosphor coatings.

devised to maintain uniform temperatures of the bulb walls for better performance of phosphor coatings. The E-bulb improves manufacturing efficiency and eliminates the clear bulb-end on phosphor-lined bulbs.

In some cases, special consideration dictate the bulb shape. The R and PAR contours have been selected to achieve desired directional distribution of light. Some of the small T-bulbs are made of highly specialized ballasts which are more economically formed in these sample contours.

The troubleshooting chart in Figure 12-2 lists troubleshooting techniques for most mercury and other HID lamp types, along with related luminaires.

Symptoms	Probable Cause	Action or Items to Check
Lamp fails to start or glows feebly.	Lamp has reached end of life.	Replace lamp.
	Lamp was too hot from previous operation.	Relight when lamp has cooled.
	Lamp loose.	Tighten in socket.
		If solder has melted or base eyelet is badly pitted, check for poor contact or defective socket.
	Low temperature.	Make sure that the ballast has adequate voltage to start the lamp at the lowest ambient temperature.

Figure 12-2: Troubleshooting chart for HID and mercury lamps and related fixtures.

Symptoms	Probable Cause	Action or Items to Check
Lamp fails to start or glows feebly.	Low voltage.	Check open-circuit voltage at socket or ballast tap if incorrectly matched to supply voltage.
		Increase line voltage if necessary.
		Look for wiring fault or poor connections.
Low light output.	Lamp is near end of life.	Replace lamp.
	Low voltage.	Check line voltage and ballast tap selected.
	Wiring fault.	Check wiring and connections.

Figure 12-2: Troubleshooting chart for HID and mercury lamps and related fixtures. *(Cont.)*

Symptoms	Probable Cause	Action or Items to Check
Low light output.	Wrong ballast.	Be sure that ballast is right for the lamp.
	Low output ballast.	Check to see if ballast delivers proper starting current.
	Excessive draft.	Protect single-tube lamps from excessive drafts.
	Dirty, corroded or inadequate fixture.	Clean, polish, or replace fixture.
	Wrong burning position.	Some fixtures are designed for operation only with their bases up or down. Observe correct burning position.

Figure 12-2: Troubleshooting chart for HID and mercury lamps and related fixtures. ***(Cont.)***

Symptoms	Probable Cause	Action or Items to Check
Lamps go out frequently for several minutes at a time.	Voltage dips.	Separate lighting circuit from heavy power circuits.
		Provide voltage regulator.
		Use ballast affording greater protection against voltage dips, such as a regulated output ballast.
	Wiring fault.	Check wiring.
		Tighten connections.
	Frequent power interruptions.	Consult utility company.
	Wrong ballast.	Install correct ballast.

Figure 12-2: Troubleshooting chart for HID and mercury lamps and related fixtures. *(Cont.)*

Symptoms	Probable Cause	Action or Items to Check
Outer bulb color abnormally greenish or yellowish. Internal parts oxidized.	Lamp is worn out from exceptionally long service, so that arc tube has discolored.	Replace lamp.
	Outer bulb seal destroyed.	Replace lamp.
	Dirt or dust.	Wipe bulb carefully so as not to scratch surface. Clean fixture.
Annoying stroboscopic effect.	Cyclic flicker.	Connect fixtures in a staggered arrangement on a three-phase supply.
		Use a two-lamp lead-lag ballast.

Figure 12-2: Troubleshooting chart for HID and mercury lamps and related fixtures. ***(Cont.)***

Symptoms	Probable Cause	Action or Items to Check
Annoying stroboscopic effect.	Cyclic flicker.	Add incandescent lamps to the system.
Radio interference.	Circuit components.	Check circuit, as lamp itself generally creates no radio interference, and standard ballast usually suppress any line re-radiation.
		If individual lamp ballasts are not used, it may be necessary to add a small capacitor across the lamp.
Noisy ballast	Cyclic vibration.	Tighten ballast cover.

Figure 12-2: Troubleshooting chart for HID and mercury lamps and related fixtures. ***(Cont.)***

Symptoms	Probable Cause	Action or Items to Check
Noisy ballast.	Cyclic vibration.	Replace ballast.
Sunburn or suntan.	Exposure to rays of lamp without outer bulb or prolonged exposure at very short distance from lamp.	Replace lamps that lack an outer bulb.
		Turn lights out when working near them or cover exposed parts of body.
		Use enclosed fixtures.
Lamp won't enter fixture.	Fixture opening too small or socket off center.	Relocate socket or try a socket extender if change in position of light center is not critical.
	Damaged base.	Replace base. Check for rough handling.

Figure 12-2: Troubleshooting chart for HID and mercury lamps and related fixtures. ***(Cont.)***

Symptoms	Probable Cause	Action or Items to Check
Lamp won't enter fixture.	Crooked base.	Check base-bulb alignment. If more than 3° in any direction, lamp may be defective.
Overcurrent devices open when lamps are started.	High transient current of very short duration. Usually caused by ballast or circuit components.	Use time-delay fused or thermal-magnetic type circuit breakers.
Lamp breakage or cracks in outer bulb.	Shipping damage or mishandling.	Check carrier and workers handling bulbs.
	Water drips on hot bulb.	Check for leaks or for condensation in fixture.
	Poorly sealed fixtures breathe in moist air.	Use enclosed vapor-tight fixture.

Figure 12-2: Troubleshooting chart for HID and mercury lamps and related fixtures. ***(Cont.)***

Symptoms	Probable Cause	Action or Items to Check
Lamp breakage.	Bulb touches fixture, edge of sockets, metallic bulb changer, heat conductor, or heat insulator during insertion or operation.	Straighten or adjust socket.
		Use socket extender to provide ample clearance.
		Remove metallic or other element touching bulb.
		Use weather duty lamps.
	Bulb cracked through mechanical shock.	Replace lamp.
		Do not allow bulbs to bump against each other.
Arc tube swollen, cracked or broken.	Overwattage operation.	Check ballast and socket wiring.

Figure 12-2: Troubleshooting chart for HID and mercury lamps and related fixtures. ***(Cont.)***

Symptoms	Probable Cause	Action or Items to Check
Arc tube swollen, cracked or broken.	Wrong burning position.	Correct burning position.
	Incorrect magnet polarity or magnet ineffective.	Correct magnet polarity or replace magnet or use lamps without magnets.
Ends of arc tube split apart or internal wires melted.	Excessive current or voltage due to operation without ballast or else lightning damage.	Check ballast and socket wiring.
		Check for possibility of lightning damage.
		Check for very high voltage applied to a hot lamp, even momentarily, resulting in discharge in outer tube.

Figure 12-2: Troubleshooting chart for HID and mercury lamps and related fixtures. ***(Cont.)***

Symptoms	Probable Cause	Action or Items to Check
Loose base.	Abnormal operation or defective lamp.	Check for operation beyond published temperature limits.
		Check if basing cement was severely disturbed prior to or within first 10 hours of operation.
Rattle.	Piece of loose glass or cement.	Lamps having this condition should not be considered defective if they are otherwise satisfactory.
Blackening of inner arc tube.	Long service.	Replace lamp if lamp output is too low.
	Overwattage operation.	Adjust.

Figure 12-2: Troubleshooting chart for HID and mercury lamps and related fixtures. ***(Cont.)***

CHAPTER 13

Troubleshooting Electric Motors

Induction motors get their name from the fact that they utilize the principle of electromagnetic induction. An induction motor has a stationary part, or stator, with windings connected to the ac supply, and a rotating part, or rotor, which contains coils or bars. There is no electrical connection between the stator and rotor. The magnetic field produced in the stator windings induces a voltage in the rotor coils or bars.

Since the stator windings act in the same way as the primary winding of a transformer, the stator of an induction motor is sometimes called the *primary*. Similarly, the rotor is called the *secondary* because it carries the induced voltage in the same way as the secondary of a transformer.

The useful life of an induction motor depends largely upon the condition of its insulation, which should be suitable for the operating requirements.

When an induction motor malfunctions, the stator (stationary) windings will usually be defective, and these windings will then have to be repaired or replaced. Stator problems can usually be traced to one or more of the following causes:

- Worn bearings

- Moisture
- Overloading
- Operating single phase
- Poor insulation

Dust and dirt are usually contributing factors. Some forms of dust are highly conductive and contribute materially to insulation breakdown. The effect of dust on the motor temperature through restriction of ventilation is another reason for keeping the machine clean, either by periodically blowing it out with compressed air or by dismantling and cleaning. The compressed air must be dry and throttled down to a low pressure that will not danger the insulation.

One of the worst enemies of motor insulation is moisture. Therefore, motor insulation must be kept reasonably dry, although many applications make this practically impossible unless a totally enclosed motor is utilized. If a motor must be operated in a damp location, special moisture-resisting treatment should be given to the windings.

The life of a winding depends upon keeping it in its original condition as long as possible. In a new machine, the winding is snug in the slots and the insulation is fresh and flexible. This condition is best maintained by periodic cleaning, followed by varnish and oven treatments.

One condition that frequently hastens winding failure is movement of the coils due to vibration during operation. After insulation dries out, it loses its flexibility and the mechanical stresses caused by starting and plugging, as well as the natural stresses in operation under load, will tend to cause short circuits in the coils and possibly failures from the coil to ground — usually at the point where the coil leaves the slot. The effect of periodic varnish and oven treatments properly carried out so as to fill all air spaces caused by drying and shrinkage of the insulation will maintain a solid winding, and also provide an effective seal against moisture.

TROUBLESHOOTING MOTORS

To detect defects in electric motors, the windings are normally tested for ground faults, opens, shorts, and reverses. The exact method of performing these tests will depend on the type of motor being serviced. However, regardless of the motor type, a knowledge of some important terms is necessary before maintenance personnel can approach their work satisfactorily.

Ground: A winding becomes grounded when it makes an electrical contact with the iron of the motor. The usual causes of grounds include the following: Bolts securing the end plates come into contact with the winding; the wires press against the laminations at the corners of the slots, which is likely to occur if the slot insulation tears or cracks during winding; and the centrifugal switch may be grounded to the end plate.

Open circuits: Loose or dirty connections as well as a broken wire can cause an open circuit in an electric motor.

Shorts: Two or more turns of the coil that contact each other electrically will cause a short circuit. This condition may develop in a new winding if the winding is tight and much pounding is necessary to place the wires in position. In other cases, excessive heat developed from overloads will make the insulation defective and will cause shorts. A short circuit is usually detected by observing smoke from the windings as the motor operates or when the motor draws excessive current at no load.

Tools for Troubleshooting

In addition to small portable testing devices such as voltmeters, ammeters, brush-spring tension testers and a transistorized stethoscope for checking motor bearings, maintenance equipment should include a 500-volt insulation-resistance tester, a spark-gap oil dielectric tester, and a portable oil-filtering unit.

Transistorized stethoscope: This type of testing instrument — equipped with a transistor-amplifier — is used to ascertain the condition of motor bearings. A little practice in interpreting what is heard through it may be required, but in general, it is relatively simple to use. If, when the stethoscope is applied to a motor bearing, a purring sound is heard, the bearing is usually normal. On the other hand, a thumping sound or a rough grinding sound indicates a failing bearing.

Insulation-resistance tests: High-voltage cables, such as those rated at 2300 volts, should be tested periodically. Circuits carrying 480 volts should be tested annually; this includes transformers, motors, motor starter, generators, and switches. Also, certain high-power process equipment such as electric furnaces and die-casting machines should receive insulation-resistance tests annually.

When performing an insulation-resistance test, first make a careful safety check and ensure that all circuits and equipment are rated at the voltage of the megger. Furthermore, all equipment scheduled for testing must be disconnected from all power sources. All safety switches should be opened and locked out to make certain that motor starters or other control equipment cannot accidentally energize the apparatus.

Before continuing, also check the megger and other testing instruments for functioning. This is accomplished by first testing for an infinity reading by operating the megger with the test leads not connected. A second test is made with the megger leads shorted and the megger handle turned very slowly; this time, the meter should read "zero."

Following these checks, insulation-resistance readings should be obtained by testing between a conductor and ground or between two conductors, or both. Insulation readings from conductor to ground are obtained by connecting the "line" test lead to the conductor and the "earth" test lead to ground. For the test between conductors, the test leads are connected to the two conductors to be tested.

After the proper connections have been made, the megger handle is turned at an even speed for about one minute. At the end of this time, the insulation-resistance value is recorded.

Because temperature and humidity have profound effects on insulation-resistance readings, temperature and humidity at the apparatus should be recorded immediately after the test. In addition, pertinent considerations should be noted such as condition of the immediate area — wet location or excessive dust, whether the apparatus has been in operation prior to the test or at rest for a prolonged period of time.

After the readings have been recorded, they are corrected for temperature using a temperature-correction chart supplied with most meggers. As a rule of thumb, most maintenance departments feel that 600-volt winding insulation is acceptable if the corrected resistance value is one megohm or more. High resistance readings which show a continuing downward trend over a period of time indicate failing insulation.

The chart in Figure 13-1 gives a list of practical tools and equipment for effective electrical maintenance, not only for motors, but for other electrical apparatus as well. Note that the tool is listed in the left column, while the application is shown in the right column.

Tools or Equipment	Application
Multimeters, voltmeters, ohmmeters, clamp-on ammeters, wattmeters, clamp-on P.F. meters	Measure circuit voltage, resistance, current and power. Useful for circuit tracing and troubleshooting.
Potential and current transformers, meter shunts	Increase range of test instruments to permit reading of high-voltage and high-current circuits.
Transistorized stethoscope	Detect faulty rotating machinery bearings; leaky valves.
Tachometer	Checks rotating machinery speeds.

Figure 13-1: Tools for effective electrical maintenance.

Tools or Equipment	Application
Recording meters, instruments	Provide permanent record of voltage, current, power, temperature, etc., on charts for analytic study.
Insulation resistance tester, thermometer, psychrometer	Test and monitor insulation resistance; use thermometer and psychrometer for temperature-humidity correction.
Portable oil dielectric tester; portable oil filter	Test OCB, transformer oil or other insulating oils. Recondition used oil.
Air gap feeler gauges	Check motor or generator air gap between rotor and stator.
Cleaning solvent	Removes grease or dirt from motor windings or other electrical parts.

Figure 13-1: Tools for effective electrical maintenance. ***(Cont.)***

Tools or Equipment	Application
Hand stones, (rough, medium, fine); grinding rig; canvas strip	Grinding, smoothing and finishing commutators or slip rings.
Spring tension scale	Checks brush pressure on dc motor commutators or on ac motor skip rings; tests electrical contact pressure on relays, starters or contactors.
Magnifying glass, binoculars	Use magnifying glass to examine brushes, commutators or small electrical contacts or parts; binoculars allow close inspection of remote or high-voltage parts.
Motor rotation tester	Checks direction of motor rotation before connection.

Figure 13-1: Tools for effective electrical maintenance. ***(Cont.)***

Grounded Coils

The usual effect of one grounded coil in a winding is the repeated blowing of a fuse, or tripping of the circuit breaker, when the line switch is closed, that is, providing the machine frame and the line are both grounded. Two or more grounds will give the same result and will also short out part of the winding in that phase in which the grounds occur. A quick and simple test to determine whether or not a ground exists in the winding can be made with a conventional continuity tester. In testing with such an instrument, first make certain that the line switch is open and locked out, causing the motor leads to be *de-energized.* Place one test lead on the frame of the motor and the other in turn on each of the line wires leading from the motor. If there is a grounded coil at any point in the winding, the lamp of the continuity tester will light, or in the case of a meter, the dial will swing toward *infinity*.

To locate the phase that is grounded, test each phase separately. In a three-phase winding it will be necessary to disconnect the star or delta connections, if accessible. After the grounded phase is located the pole-group connections in that phase can be disconnected and each group tested separately. When the leads are placed, one on the frame and the other on the grounded coil group, the lamp will indicate the ground in this group by lighting. The stub connections between the coils and this group may then be disconnected and each coil tested separately until the exact coil that is grounded is located.

Sometimes moisture in the insulation around the coils on old and defective insulation will cause a high-resistance ground that is difficult to detect with a test lamp. A megger can be used to detect such faults, but in many cases a megger may not be available. If not, use a test outfit consisting of a headphone set (telephone receiver) and several dry cell batteries connected in series as shown in Figure 13-2. Such a test set will detect a ground of very high resistance, and this set will often be found very effective when the ordinary test lamp fails to locate the trouble.

Armature windings and the commutator of a motor may be tested for grounds in a similar manner. On some motors, the brush holders are grounded to the end plate.

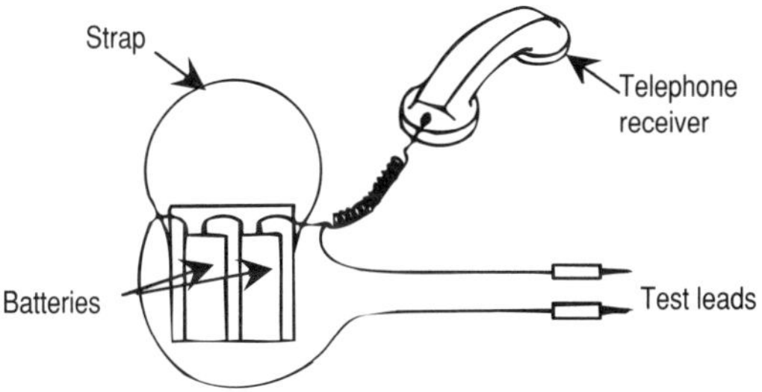

Figure 13-2: When used for testing, a clicking sound indicates a fault.

Consequently, before the armature is tested for grounds, the brushes must be lifted away from the commutator.

When a grounded coil is located, it should be either removed and reinsulated or cut out of the circuit. At times, however, it may be inconvenient to stop a motor long enough for a complete rewinding or permanent repairs. In such cases, when trouble develops, it is often necessary to make a temporary repair until a later time when the motor may be taken out of service long enough for rewinding or permanent repairs.

To temporarily repair a defective coil, a jumper wire of the same size as that used in the coils is connected to the bottom lead of the coil immediately adjacent to the defective coil and run across to the top lead of the coil on the other side of the defective coil, leaving the defective coil entirely out of the circuit. The defective coil should then be cut at the back of the winding and the leads taped so as not to function when the motor is started again. If the defective coil is grounded, it should also be disconnected from the other coils.

Shorted Coils

Shorted turns within coils are usually the result of failure of the insulation on the wires. This is frequently caused by the wires being crossed and having excessive pressure applied on the crossed conductors when the coils are being inserted in the slot. Quite often it is caused by using too much force in driving the coils down in the slots. In the case of windings that have been in service for several years, failure of the insulation may be caused by oil, moisture, etc. If a shorted coil is left in a winding, it will usually burn out in a short time and if it is not located and repaired promptly will probably cause a ground and the burning out of a number of other coils.

One inexpensive way of locating a shorted coil is by the use of a growler and a thin piece of steel. Figure 13-3 shows a sketch of a growler in use in a stator. Note that the poles are shaped to fit the curvature of the teeth inside the stator core. The growler should be placed in the core

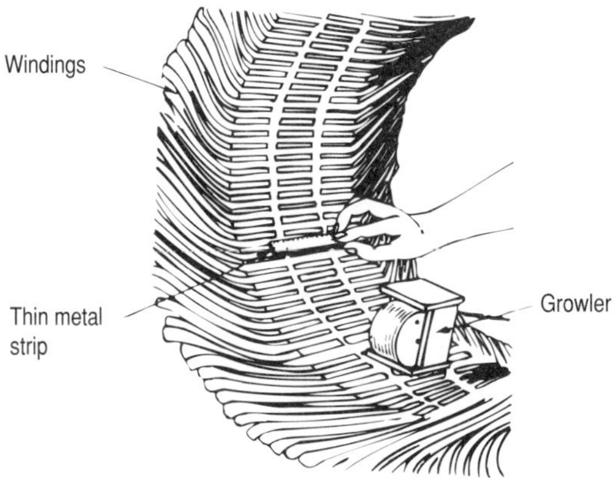

Figure 13-3: Growler used to test a stator of an ac motor.

as shown, and the thin piece of steel should be placed the distance of one coil span away from the center of the growler. Then, by moving the growler around the bore of the stator and always keeping the steel strip the same distance away from it, all of the coils can be tested.

If any of the coils has one or more shorted turns, the piece of steel will vibrate very rapidly and cause a loud humming noise. By locating the two slots over which the steel vibrates, both sides of the shorted coil can be found. If more than two slots cause the steel to vibrate, they should all be marked, and all shorted coils should be removed and replaced with new ones or cut out of the circuit as previously described.

Sometimes one coil or a complete coil group becomes short-circuited at the end connections. The test for this fault is the same as that for a shorted coil. If all the coils in one group are shorted, it will generally be indicated by the vibration of the steel strip over several consecutive slots, corresponding to the number of coils in the group.

The end connections should be carefully examined, and those that appear to have poor insulation should be moved during the time that the test is being made. It will often be found that when the shorted end connections are moved during the test the vibration of the steel will stop. If these ends are reinsulated, the trouble should be eliminated.

Open Coils

When one or more coils become open-circuited by a break in the turns or a poor connection at the end, they can be tested with a continuity tester as previously explained. If this test is made at the ends of each winding, an open can be detected by the lamp failing to light. The insulation should be removed from the pole-group connections, and each group should be tested separately.

An open circuit in the starting winding may be difficult to locate, since the problem may be in the centrifugal switch as well as the winding itself. In fact, the centrifugal switch is probably more apt to cause trouble than the winding since parts become worn, defective, and more likely, dirty. Insufficient pressure of the rotating part of centrifugal switches against the stationary part will prevent the contacts from closing and thereby produce an open circuit.

If the trouble is a loose connection at the coil ends, it can be repaired by resoldering the splices, but if it is within the coil, the coil should either be replaced or a jumper should be connected around it until a better repair can be made.

Reversed Connections

Reversed coils cause the current to flow through them in the wrong direction. This fault usually manifests itself — as do most irregularities in winding connections

— by a disturbance of the magnetic circuit, which results in excessive noise and vibration. The fault can be located by the use of a magnetic compass and some source of low-voltage direct current. This voltage should be adjusted so it will send about one-fourth to one-sixth of the full-load current through the winding, and the dc leads should be placed on the start and finish of one phase. If the winding is three-phase, star-connected, this would be at the start of one phase and the star point. If the winding is delta-connected, the delta must be disconnected and each phase tested separately.

Place a compass on the inside of the stator and test each of the coil groups in that phase. If the phase is connected correctly, the needle of the compass will reverse definitely as it is moved from one coil group to another. However, if any one of the coils is reversed, the reversed coil will build up a field in the direction opposite to the others, thus causing a neutralizing effect which will be indicated by the compass needle refusing to point definitely to that group. If there are only two coils per group, there will be no indication if one of them is reversed, as that group will be completely neutralized.

When an entire coil group is reversed, it causes the current to flow in the wrong direction in the whole group. The test for this fault is the same as that for reversed coils. The winding should be magnetized with direct current, and when the compass needle is passed around the coil groups, they should indicate alternately N.S.,

N.S., etc. If one of the groups is reversed, three consecutive groups will be of the same polarity. The remedy for either reversed coil groups or reversed coils is to make a visual check of the connections at that part of the winding, locate the wrong connection, and reconnect it properly.

When the wrong number of coils are connected in two or more groups, the trouble can be located by counting the number of ends on each group. If any mistakes are found, they should be remedied by reconnecting properly.

Reversed Phase

Sometimes in a three-phase winding a complete phase is reversed by either having taken the starts from the wrong coils or by connecting one of the windings in the wrong relation to the others when making the star or delta connections. If the winding is delta connected, disconnect any one of the points where the phases are connected together and pass current through the three windings in series. Place a compass on the inside of the stator and test each coil group by slowly moving the compass one complete revolution around the stator.

The reversals of the needle in moving the compass one revolution around the stator should be three times the number of poles in the winding.

In testing a star- or wye-connected winding, connect the three starts together and place them on one dc lead. Then connect the other dc lead and star point, thus passing the current through all three windings in parallel. Test with a compass as explained for the delta winding. The result should then be the same, or the reversals of the needle in making one revolution around the stator should again be three times the number of poles in the winding.

These tests for reversed phases apply to full-pitch windings only. If the winding is fractional pitch, a careful visual check should be made to determine whether there is a reversed phase or mistake in connecting the star or delta connections.

The general troubleshooting chart in Figure 13-4 may be used by qualified personnel who have the proper tools and equipment. These instructions do not cover all details or variations in equipment, nor do they provide for every possible condition to be met in actual practice.

TROUBLESHOOTING SPLIT-PHASE MOTORS

If a split-phase motor fails to start, the trouble may be due to one or more of the following faults:

- Tight or "frozen" bearings

Malfunction	Probable Cause	Corrective Action
Slow speed	Open primary circuit.	Locate fault with testing device and repair.
Slow to accelerate	Excess loading.	Reduce load.
	Poor circuit.	Check for high resistance.
	Defective squirrel-cage rotor.	Replace.
	Applied voltage too low.	Get power company to increase voltage tap.
Wrong rotation	Wrong sequence of phases.	Reverse connections at motor or at switchboard.

Figure 13-4: General troubleshooting chart for motors.

Malfunction	Probable Cause	Corrective Action
Motor overheats	Check for overload.	Reduce load.
	Wrong blowers or air shields.	May be clogged with dirt and prevent proper ventilation of motor.
	Motor may have one phase open.	Check to make sure that all leads are well connected.
	Grounded coil.	Locate and repair.
	Unbalanced terminal voltage.	Check to make sure that all leads are well connected.
	Grounded coil.	Locate and repair.

Figure 13-4: General troubleshooting chart for motors. ***(Cont.)***

Malfunction	Probable Cause	Corrective Action
Motor overheats	Unbalanced terminal voltage.	Check for faulty leads.
	Shorted stator coil.	Repair and then check wattmeter reading.
	Faulty connection.	Indicate by high resistance.
	High voltage.	Check terminals of motor with voltmeter.
	Low voltage.	Same as above.
Motor stalls	Wrong application.	Change type or size. Consult manufacturer.

Figure 13-4: General troubleshooting chart for motors. ***(Cont.)***

Malfunction	Probable Cause	Corrective Action
Motor stalls	Overloaded motor.	Reduce load.
	Low motor voltage.	See that nameplate voltage is maintained.
	Open Circuit.	Fuses blown.
	Incorrect control resistance of wound rotor.	Check control sequence. Replace broken resistors. Repair open circuits.
Motor does not start	One phase open.	See that no phase is open. Reduce load.
	Defective rotor.	Look for broken bars or rings.

Figure 13-4: General troubleshooting chart for motors. ***(Cont.)***

Malfunction	Probable Cause	Corrective Action
Motor does not start	Poor stator coil connection.	Remove end bells.
Motor runs, then quits	Power failure.	Check for loose connections to line, to fuses and to control.
Slow speed	Not applied properly.	Consult supplier for proper type.
	Voltage too low at motor terminals because of line drop.	Use higher voltage on transformer terminals or reduce load.

Figure 13-4: General troubleshooting chart for motors. *(Cont.)*

Malfunction	Probable Cause	Corrective Action
Slow speed	If wound rotor, improper control operation of secondary.	Correct secondary control.
	Low pull-in torque of synchronous motor.	Change rotor starting resistance or change rotor design.
	Check that all brushes are riding on rings.	Check secondary connections; leave no leads poorly connected.
	Broken rotor bars.	Look for cracks near the rings. A new rotor may be required.

Figure 13-4: General troubleshooting chart for motors. ***(Cont.)***

Malfunction	Probable Cause	Corrective Action
Motor vibrates	Motor misaligned.	Realign.
	Weak foundations.	Strengthen base.
	Coupling out of balance.	Balance coupling.
	Driven equipment unbalanced.	Rebalance driven equipment.
	Defective ball bearing.	Replace bearing.
	Bearing not in line.	Line up properly.
	Balancing weights shifted.	Rebalance motor.

Figure 13-4: General troubleshooting chart for motors. ***(Cont.)***

Malfunction	Probable Cause	Corrective Action
Motor vibrates	Wound rotor coils replaced.	Rebalance motor.
	Polyphase motor running single-phase.	Check for open circuit.
	Excessive end play.	Adjust bearing or add washer.
Unbalanced line current	Unequal terminal volts.	Check leads and connections.
	Single-phase operation.	Check for open circuit.
	Poor rotor contacts in control wound rotor resistance.	Check control devices.

Figure 13-4: General troubleshooting chart for motors. ***(Cont.)***

Malfunction	Probable Cause	Corrective Action
Unbalanced line current	Brushes not in proper posiiont in wound rotor.	See that brushes are properly seated and shunts in good condition.
	Fan rubbing air shield.	Remove interference.
	Fan striking insulation.	Clear fan.
	Loose bedplate.	Tighten holding bolts.
Magnetic noise	Air gap not uniform.	Check and correct bracket fits or bearing.

Figure 13-4: General troubleshooting chart for motors. *(Cont.)*

- Worn bearings, allowing the rotor to drag on the stator
- Bent rotor shaft
- One or both bearings out of alignment
- Open circuit in either starting or running windings
- Defective centrifugal switch
- Improper connections in either winding
- Grounds in either winding or both
- Shorts between the two windings

Tight or worn bearings: Tight or worn bearings may be due to the lubricating system failing, or when new bearings are installed, they may run hot if the shaft is not kept well oiled.

If the bearings are worn to such an extent that they allow the rotor to drag on the stator, this will usually prevent the rotor from starting. The inside of the stator laminations will be worn bright where they are rubbed by the rotor. When this condition exists, it can generally be easily detected by close observation of the stator field and rotor surface when the rotor is removed.

Bent shaft and bearings out of line: A bent rotor shaft will usually cause the rotor to bind when in a certain position and then run freely until it comes back to the same position again. An accurate test for a bent shaft can be made by placing the rotor between centers on a lathe and turning the rotor slowly while a tool or marker is held in the tool post close to the surface of the rotor. If the rotor wobbles, it is an indication of a bent shaft.

Bearings out of alignment are usually caused by uneven tightening of the end-shield plates. When placing end shields or brackets on a motor, the bolts should be tightened alternately, first drawing up two bolts which are diametrically opposite. These two should be drawn up only a few turns and the other kept tightened an equal amount all the way around. When the end shields are drawn up as far as possible with the bolts, they should be tapped tightly against the frame with a mallet and the bolts tightened again.

Open circuits and defective centrifugal switches: Open circuits in either the starting or running winding will cause the motor to fail to start. This fault can be detected by testing in series with the start and finish of each winding with a test lamp or ohmmeter.

A defective centrifugal switch will often cause considerable trouble that is difficult to locate unless one has good knowledge of the operating characteristics of these switches. If the switch fails to close when the rotor stops, the motor will not start when the line switch is closed. Failure of the switch to close is generally caused by dirt, grit, or some other foreign matter getting into the switch. The switch should be thoroughly cleaned with a degreasing solution such as AWA1,1,1 and then inspected for weak or broken springs.

If the winding is on the rotor, the brushes sometimes stick in the holders and fail to make good contact with the slip rings. This causes sparking at the brushes. There will probably also be a certain place where the rotor will not start until it is moved far enough for the brush to make contact on the ring. The brush holders should be cleaned and the brushes carefully fitted so they move more freely with a minimum of friction between the brush and the holders. If a centrifugal switch fails to open when the motor is started, the motor will probably growl and continue to run slowly, causing the starting winding to burn out if not promptly disconnected from the line. In most cases, however, the "heaters" in the motor control will take care of this before any serious damage occurs. This fault is likely to be caused by dirt or hardened grease in the switch.

Reversed connections and grounds: Reversed connections are caused by improperly connecting a coil or group of coils. The wrong connections can be found and corrected by making a careful check on the connections and reconnecting those that are found at fault. The test with a dc power source and a compass can also be used for locating reversed coils. Test the starting and running windings separately, exciting only one winding at a time, with direct current. The compass should show alternate poles around the winding.

The operation of a motor that has a ground in the winding will depend on where the ground is and whether or not the frame is grounded. If the frame is grounded, then when the ground occurs in the winding, it will usually blow a fuse or trip the overcurrent protective device.

A test for grounds can be made with a test lamp or continuity tester. One test lead should be placed on the frame and the other on a lead to the winding. If there is no ground, the lamp will not light, nor will any deflection be present when a meter is used. If the light does light or the meter shows continuity, it indicates a ground is present — due to a defect somewhere in the motor's insulation.

Short circuits: Short circuits between any two windings can be detected by the use of a test lamp or continuity tester. Place one of the test leads on one wire of the starting winding and the other test lead on the wire of the running winding. If these windings are properly insulated from each other, the lamp should not light. If it does, it is a certain indication that a short or ground fault exists between the windings. Such a condition will usually cause part of the starting winding to burn out. The starting winding is always wound on top of the running winding, so if it becomes defective due to a defective centrifugal switch or a short circuit, the starting winding can be conveniently removed and replaced without disturbing the running winding.

STORING MOTORS

There are many reasons for storing motors, but the two major ones are:

- The project on which they are to be used is not complete.
- Spare motors are often kept as back-ups on most industrial installations.

The first consideration when storing motors for any length of time is the location. A dry location should be selected if at all possible — one that does not undergo severe changes in temperature over a 24-hour period. When the ambient temperature changes frequently during a 24-hour period, condensation is certain to form on the motor, and moisture is one of the worst enemies of motor insulation. Therefore, guarding against moisture is one of the chief concerns when storing motors of any type.

A means for transporting the motors from the place of storage to the place where it will be used, or else shifted around in the storage area, is also of importance. Motors should *not* be lifted by its rotating shaft. Doing so can damage the alignment of the rotor in relationship to the stator. Even picking up the smaller fractional horsepower motors by the shaft is not recommended. Many workers have received bad cuts from the sharp keyways on motor shafts when picked up with the bare hands.

CAUTION

Never lift a motor by its shaft; especially ones that can be lifted manually. The sharp edges of keyways are not unlike sharp kitchen knives which can cause deep cuts in the body.

When an electric motor is received at the job site, always refer to the manufacturer's instructions and follow them to the letter. Failure to do so could result in serious injury to the workers and motor alike.

Once the motor has been uncrated, check to see if any damage has occurred during handling. Be sure that the motor shaft and armature turn freely. This is also a good time to check to determine if the motor has been exposed to dirt, grease, grit, or excessive moisture in either shipment of storage. Motors in storage should have shafts turned over once each month to redistribute grease in the bearings.

Note: *Motors in storage should have shafts turned over once each month to redistribute grease in the bearings.*

WARNING!

Never start a motor which has been wet without first having it thoroughly dried.

The measure of insulation resistance is a good dampness test. Clean the motor of any dirt or grit before putting it back in service.

Eyebolts of lifting lugs on motors are intended only for lifting the motor and factory motor-mounted standard accessories. These lifting provisions should never be used when lifting or handling the motor when the motor is attached to other equipment as a single unit.

The eyebolt lifting-capacity rating is based on a lifting alignment coincident with the eyebolt centerline. The eyebolt capacity reduces as deviation from this alignment increases.

The following is a list of items that must be considered when storing motors for any length of time:

- Make sure motors are kept clean.
- Make sure motors are kept dry.
- Supply supplemental heating in the storage area if necessary.
- Motors should be stored in an orderly fashion; that is, grouped by horsepower, etc.
- Motor armatures should be rotated periodically.
- Lubrication should be checked periodically.
- Protect shafts and keyways during storage and also while transporting motors from one location to another.

- Test motor-winding resistance upon receiving; test again after setting in storage.

IDENTIFYING MOTORS

Electrical workers will sometimes come across a motor with no identification (no nameplate or lead tags) which must be put back into service or else repaired. The experienced electrician should know how to positively identify the motor's characteristics, even with no written data.

The NEMA Standard method of motor identification is easy to remember by drawing the coils to form a wye. Identify one outside coil end with the Number one (1), and then draw a decreasing spiral and number each coil end in sequence as shown in Figure 13-5.

By using an ohmmeter or other continuity tester, the individual circuits can be located as follows.

Step 1. Connect one probe of the tester to any lead, and check for continuity to each of the other 8 leads. A reading from only one other lead indicates one of the two wire circuits. A reading to two other leads indicates the three wire circuit that makes up the internal wye connection.

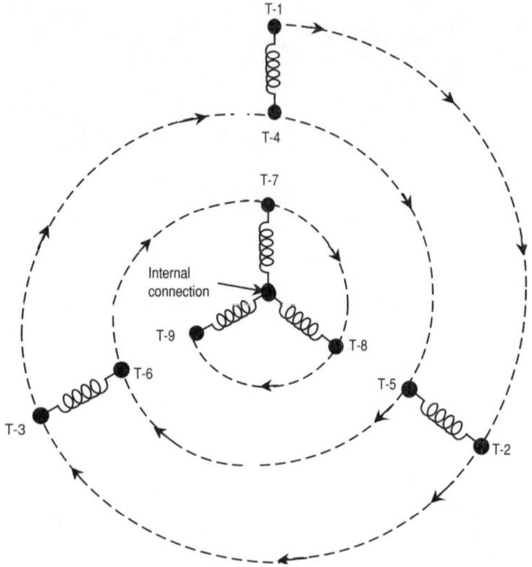

Figure 13-5: Identify one outside coil and then draw a decreasing spiral and number each coil.

Step 2. Continue checking and isolating leads until all four circuits have been located. Tag the wires of the three lead circuits T-7, T-8, and T-9 in any order. The other leads should be temporarily marked T-1 and T-4 for one circuit, T-2 and T-5 for the second circuit, T-3 and T-6 for the third and final circuit.

The following test voltages are for the most common dual voltage range of 230/460 volts. For other motor ranges, the voltages listed should be changed in proportion to the motor rating.

As all the coils are physically mounted in slots on the same motor frame, the coils will act almost like the primary and secondary coils of a transformer. Figure 13-6 shows a simplified electrical arrangement of the coils. Depending on which coil group power is applied to, the resulting voltage readings will be additive, subtractive, balanced, or unbalanced depending on physical location with regard to the coils themselves.

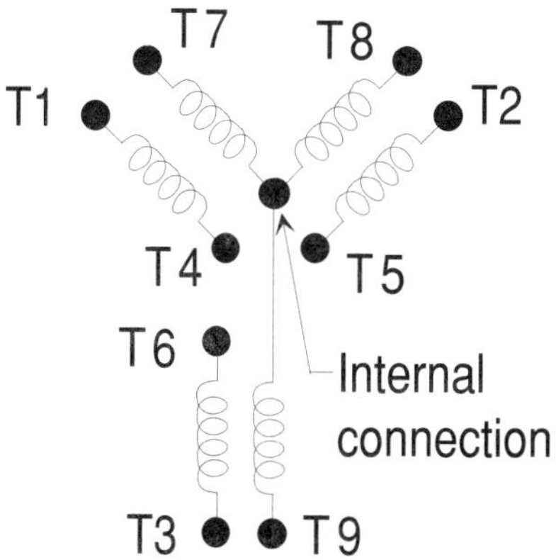

Figure 13-6: Simplified electrical arrangement of wye-wound motor coils.

Step 3. The motor may be started on 230 volts by connecting leads T-7, T-8, and T-9 to the three phase source. If the motor is too large to be connected directly to the line, the voltage should be reduced by using a reduced voltage starter or other suitable means.

Step 4. Start the motor with no load connected and bring up to normal speed.

Step 5. With the motor running, a voltage will be induced in each of the open two-wire circuits that were tagged T-1 and T-4, T-2 and T-5, T-3 and T-6. With a voltmeter, check the voltage reading of each circuit. The voltage should be approximately 125-130 volts and should be the same on each circuit.

Note: The voltages referred to during the testing are only for reference and will vary greatly from motor to motor, depending on size, design, and manufacturer. If the test calls for equal voltages of 125-130 and the reading is only 80-90, that is okay as long as the voltage readings are nearly equal.

Step 6. With the motor still running, carefully connect the lead that was temporarily marked T-4 with the T-7 and line lead. Read the voltage between T-1 and T-8 and also between T-1 and T-9. If both readings are of the same value and are approximately 330-340 volts, leads T-1 and T-4 may be disconnected and permanently marked T-1 and T-4.

Step 7. If the two voltage readings are of the same value and are approximately 125-130, disconnect and interchange leads T-1 and T-4 and mark permanently (original T-1 changed to T-4 and original T-4 changed to T-1).

Step 8. If readings between T-1 and T-8 and between T-1 and T-9 are of unequal values, disconnect T-4 from T-7 and reconnect T-4 to the junction of T-8 and line.

Step 9. Measure the voltage now between T-1 and T-7 and also between T-1 and T-9. If the voltages are equal and approximately 330-340 volts, tag T-1 is permanently marked T-2 and T-4 is marked T-5 and disconnected. If the readings taken are equal but are approximately 125-130 volts, leads T-1 and T-4 are disconnected, interchanged, and marked T-2 and T-5 (T-1 changed to T-5, and T-4 changed to T-2). If both voltage readings are different, T-4 lead is disconnected from T-8 and moved to T-9. Voltage readings are taken again (between T-1 and T-7, T-1 and T-8) and the leads permanently marked T-3 and T-6 when equal readings of approximately 330-340 volts are obtained.

Step 10. The same procedure is followed for the other two circuits that were temporarily marked T-2 and T-5, and T-3 and T-6 until a position is found where both voltage readings are equal and approximately 330-340 volts and the tags change to correspond to the standard lead markings as shown in Figure 13-7.

Step 11. Once all leads have been properly and permanently tagged, leads T-4, T-5, and T-6 are connected together and voltage readings are taken between T-1, T-2, and T-3. The voltages should be equal and approximately 230 volts.

Step 12. As an additional check, the motor is shut down and leads T-7, T-8, and T-9 are disconnected, and leads T-1, T-2, and T-3 are connected to the line. Connect T-1 to the line lead T-7 was connected to, T-2 to the same line as T-8 was previously connected, and T-3 to the same lead that T-9 was connected to. With T-4, T-5, and T-6 still connected together to form a wye connection, the motor can again be started without a load. If all lead markings are correct, the motor rotation with leads T-1, T-2, and T-3 connected will be the same as when T-7, T-8, and T-9 were connected.

The motor is now ready for service and is connected in series for high voltage or parallel for low as indicated by the NEMA Standard connections shown in Figure 13-7.

Note: This procedure may not work on some wye-wound motors with concentric coils.

Three-Phase Delta-Wound Motors

Most dual-voltage, delta-wound motors also have 9 leads, as indicated in Figure 13-8, but there are only three circuits of three leads each.

Continuity tests are used to find the three coil groups as was done for the wye-wound motor. Once the coil groups are located and isolated, further resistance checks must be made to locate the common wire in each coil

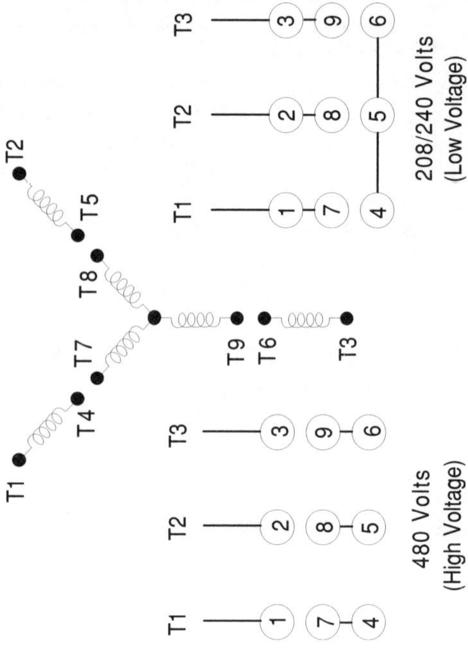

Figure 13-7: NEMA Standard lead markings for dual-voltage, wye-wound motors.

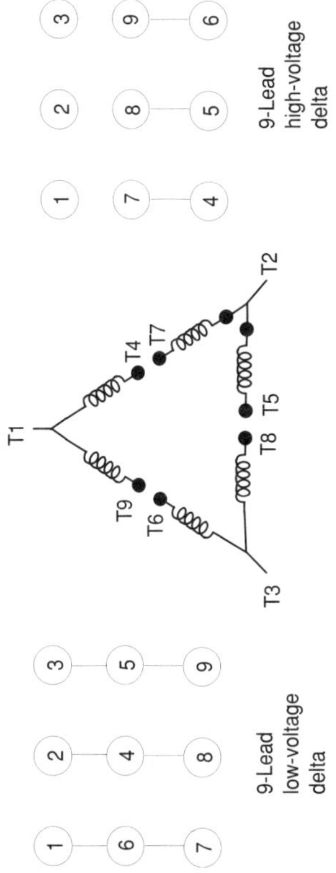

Figure 13-8: NEMA Standard lead markings for dual-voltage, delta-wound motors.

group. As the resistance of some delta wound motors is VERY low, a digital ohmmeter, wheatstone bridge, or other sensitive device may be needed.

Each coil group consists of two coils tied together with three leads brought out to the motor junction or terminal box. Reading the resistance carefully between each of the three leads shows that the readings from one of the leads to each of the other two leads will be the same (equal), but the resistance reading between those two leads will be double the previous readings; Figure 13-9 may help clarify the technique.

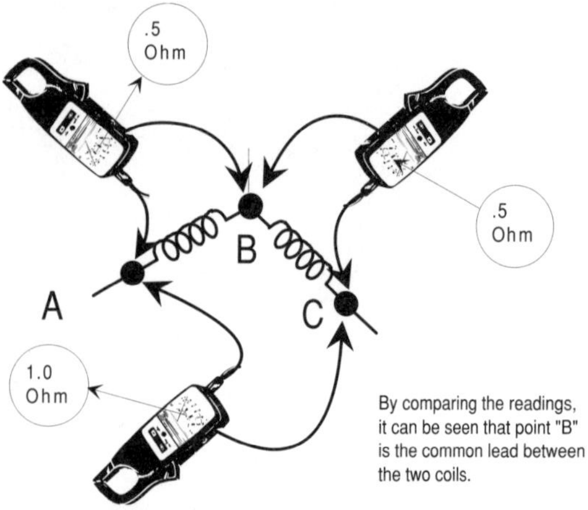

Figure 13-9: Using ohmmeter to test motor leads.

The common lead found in the first coil group is permanently marked T-1, and the other two leads temporarily marked T-4 and T-9. The common lead of the next coil group is found and permanently marked T-2 and the other leads temporarily marked T-5 and T-7. The common lead of the last coil group is located and marked T-3 with the other leads being temporarily marked T-6 and T-8.

After the leads have been marked, the motor may be connected to a 230 V three-phase line using leads T-1, T-4, and T-9. Lead T-7 is connected to line and T-4, and the motor is started with no load connected. Voltage readings are taken between T-1 and T-2. If the voltage is approximately 460, the markings are correct and may be permanently marked.

If the voltage reading is 400 volts or less, interchange T-5 and T-7 OR T-4 and T-9 and read the voltage again. If the voltage is approximately 230 volts, interchange BOTH T-5 with T-7 and T-4 with T-9. The readings should now be approximately 460 between leads T-1 and T-2. The leads connected together now are actually T-4 and T-7 and are marked permanently. The remaining lead in each group can now be marked T-9 and T-5 as indicated by Figure 13-8.

Connect one of the leads of the last coil group (not T-3) to T-9. If the reading is approximately 460 V between T-1 and T-3, the lead may be permanently marked T-6. If the reading is 400 volts or less, interchange T-6 and T-8. A reading now of 460 volts should exist between T-1 and T-3. T-6 is changed to T-8 and marked permanently and temporary T-8 is changed to T-6.

If all leads are now correctly marked, equal readings of approximately 460 volts can be obtained between leads T-1, T-2, and T-3.

To double check the markings, the motor is shut off and reconnected using T-2, T-5, and T-7. T-2 is connected to the same line lead as T-1, lead T-5 is connected where T-4 was, and T-7 is hooked where T-9 was previously connected. When started, the motor should rotate the same direction as before.

Stop the motor and connect leads T-3, T-6, and T-8 to the line leads previously connected to T-2, T-5, and T-7, respectively, and when the motor is started it should still rotate in the same direction.

The motor is now ready for service and is connected in series for high or parallel for low voltage as indicated by the NEMA Standard connections shown in Figure 13-8.

Summary

The first step toward a reliable maintenance program is to prepare records. And, while this chore often is distasteful, it also is essential to tabulate the mass of pertinent data which cannot possibly be retained by memory alone. Obviously, records can take many forms and must be tailored to specific needs and resources. But, as a minimum, records on each motor should include:

- A complete description, including age and nameplate data.
- Location and application, keeping such notations up-to-date if motors are transferred to different areas or used for different purposes.
- Notations of scheduled preventive maintenance and previous repair work performed.
- Location of duplicate or interchangeable motors.
- An estimate of the motor's importance in the productive process to which it relates.

In determining which motors are likely to fail first, it is well to remember that motor failures generally are caused by either loading, age, vibration, contamination, or commutation problems.

Troubleshooting Charts

The troubleshooting charts that follow give most motor problems found in the industry, along with their causes and remedies.

Symptoms	Probable Cause	Action or Items to Check
Hot Bearings — General	Bent or sprung shaft.	Straighten or replace shaft.
	Excessive belt pull.	Decrease belt tension.
	Pulley too far away.	Move pulley closer to bearing.
	Pulley diameter too small.	Use larger pulleys.
	Misalignment.	Correct by realignment of drive.

Figure 13-10: Troubleshooting chart for electric motors.

Symptoms	Probable Cause	Action or Items to Check
Hot Bearings — Sleeve	Oil grooving in bearing obstructed by dirt.	Remove bracket or pedestal with bearing and clean oil grooves and bearing housing; renew oil.
	Bent or damaged oil rings.	Repair or replace oil rings.
	Oil too heavy.	Use a recommended lighter oil.
	Oil too light.	Use a recommended heavier oil.

Figure 13-10: Troubleshooting chart for electric motors. ***(Cont.)***

Symptoms	Probable Cause	Action or Items to Check
Hot Bearings — Sleeve	Insufficient oil.	Fill reservoir to proper level in overflow plug with motor at rest.
	Too much end thrust.	Reduce thrust induced by driven machine or supply external means to carry thrust.
	Badly worn bearing.	Replace bearing.
Hot Bearings — Ball	Insufficient grease.	Replace bearing.

Figure 13-10: Troubleshooting chart for electric motors. ***(Cont.)***

Symptoms	Probable Cause	Action or Items to Check
Hot Bearings — Ball	Deterioration of grease or lubricant contaminated.	Remove old grease, wash bearings thoroughly in kerosene and replace with new grease.
	Excess lubricant.	Reduce quantity of grease. Bearing should not be more than ½ filled.
	Heat from hot motor or external source.	Protect bearing by reducing motor temperature.

Figure 13-10: Troubleshooting chart for electric motors. ***(Cont.)***

Symptoms	Probable Cause	Action or Items to Check
Hot Bearings — Ball	Overloaded bearing.	Check alignment, side thrust and end thrust.
	Broken ball or rough races.	Replace bearing; first clean housing thoroughly.
Oil leakage from overflow plugs	Stem of overflow plug not tight.	Remove, recement threads, replace and tighten.

Figure 13-10: Troubleshooting chart for electric motors. ***(Cont.)***

Symptoms	Probable Cause	Action or Items to Check
Oil leakage from overflow plugs	Cracked or broken overflow plug.	Replace the plug.
	Plug cover not tight.	Requires cork gasket, or if screw type, may be tightened.
Motor dirty	Ventilation blocked, end windings filled with fine dust or lint.	Clean motor will run 10° to 30° C cooler. Dust may be cement, sawdust, rock dust, grain dust, coal dust and the like. Dismantle entire motor and clean all windings and parts.

Figure 13-10: Troubleshooting chart for electric motors. ***(Cont.)***

Symptoms	Probable Cause	Action or Items to Check
Motor dirty	Rotor winding clogged.	Clean, grind and undercut commutator. Clean and treat windings with good insulating varnish.
	Bearing and brackets coated inside.	Dust and wash with cleaning solvent.
Motor wet	Drenched condition.	Motor should be covered to retain heat and the rotor position shifted frequently.

Figure 13-10: Troubleshooting chart for electric motors. ***(Cont.)***

Symptoms	Probable Cause	Action or Items to Check
Motor wet	Subject to dripping.	Wipe motor and dry by circulating heated air through motor. Install drip or canopy type covers over motor for protection.
	Submerged in flood waters.	Dismantle and clean parts. Bake windings in oven at 105° C for 24 hours or until resistance to ground is sufficient. First make sure commutator bushing is drained of water.

Figure 13-10: Troubleshooting chart for electric motors. ***(Cont.)***

Symptoms	Probable Cause	Action or Items to Check
Fails to start	Circuit not complete.	Switch open, leads broken.
	Brushes not down on commutator.	Held up by brush springs, need replacement. Brushes worn out.
	Brushes stuck in holders.	Remove and sand, clean up brush boxes.
	Armature locked by frozen bearings in motor or main drive.	Remove brackets and replace bearings or recondition old bearings if inspection makes possible.

Figure 13-10: Troubleshooting chart for electric motors. ***(Cont.)***

Symptoms	Probable Cause	Action or Items to Check
Fails to start	Power may be off.	Check line connections to starter with light. Check contacts in starter.
Motor starts, then stops and reverses direction of rotation	Reverse polarity of generator that supplies power.	Check generating unit for cause of changing polarity.

Figure 13-10: Troubleshooting chart for electric motors. ***(Cont.)***

Symptoms	Probable Cause	Action or Items to Check
Motor starts, then stops and reverses direction of rotation	Shunt and series fields are bucking each other.	Reconnect either the shunt or series field so as to correct the polarity. Then connect armature leads for desired direction of rotation. The fields can be tried separately to determine the direction of rotation individually.
Motor does not come up to rated speed	Overload.	Check bearing to see if in first class condition with correct lubrication. Check driven load for excessive load or friction.

Figure 13-10: Troubleshooting chart for electric motors. ***(Cont.)***

Symptoms	Probable Cause	Action or Items to Check
Motor does not come up to rated speed	Starting resistance not all out.	Check starter to see if mechanically and electrically incorrect.
	Voltage low.	Measure voltage with meter and check with motor nameplate.
	Short circuit in armature windings or between bars.	For shorted armature inspect commutator for blackened bars and burned adjacent bars. Inspect windings for burned coils or wedges.

Figure 13-10: Troubleshooting chart for electric motors. ***(Cont.)***

Symptoms	Probable Cause	Action or Items to Check
Motor does not come up to rated speed	Starting heavy load with very weak field.	Check full field relay and possibilities of full field setting of the field rheostat.
	Motor off neutral.	Check for factory setting of brush rigging or test motor for true neutral setting.
	Motor cold.	Increase load on motor so as to increase its temperature, or add field rheostat to set speed.

Figure 13-10: Troubleshooting chart for electric motors. ***(Cont.)***

Symptoms	Probable Cause	Action or Items to Check
Motor runs too fast	Voltage above rated.	Correct voltage or get recommended change in air gap from manufacturer.
	Load too light.	Increase load or install fixed resistance in armature circuit.
	Shunt field coil shorted.	Install new coil.
	Shunt field coil reversed.	Reconnect coil leads in reverse.

Figure 13-10: Troubleshooting chart for electric motors. ***(Cont.)***

Symptoms	Probable Cause	Action or Items to Check
Motor runs too fast	Series coil reversed.	Reconnect coil leads in reverse.
	Series field coil shorted.	Install new or repaired coil.
	Neutral setting shifted off neutral.	Reset neutral by checking factory setting mark or testing for neutral.
	Part of shunt field rheostat or unnecessary resistance in field circuit.	Measure voltage across field and check with nameplate rating.

Figure 13-10: Troubleshooting chart for electric motors. ***(Cont.)***

Symptoms	Probable Cause	Action or Items to Check
Motor runs too fast	Motor ventilation restricted causing hot shunt field.	Hot field is high in resistance; check causes for hot field in order to restore normal shunt field current.
Motor gaining speed steadily and increasing load does not slow it down	Unstable speed load regulation.	Inspect motor to see if off neutral. If series field has a shunt around the series circuit that can be removed, check series field to determine shorted turns.

Figure 13-10: Troubleshooting chart for electric motors. ***(Cont.)***

Symptoms	Probable Cause	Action or Items to Check
Motor gaining speed steadily and increasing load does not slow it down	Reversed field coil shunt or series.	Test with compass and reconnect coil.
	Too strong a commutating pole or commutating pole air gap too small.	Check with factory for recommended change in coils or air gap.
Motor runs too slow continuously	Voltage below rated.	Check voltage and correct to value on nameplate.
	Motor operates cold.	Motor may run 20% slow due to light load. Install smaller motor.

Figure 13-10: Troubleshooting chart for electric motors. ***(Cont.)***

Symptoms	Probable Cause	Action or Items to Check
Motor runs too slow continuously	Overload.	Check bearings of motors and the drive to see if in first class condition. Check for excessive friction in drive.
	Neutral setting shifted.	Check for factory setting of brush rigging or test for true neutral setting.
	Armature has shorted coils or commutator bars.	Remove armature to repair shop and put in first class condition.

Figure 13-10: Troubleshooting chart for electric motors. ***(Cont.)***

Symptoms	Probable Cause	Action or Items to Check
Motor overheats or runs hot	Overloaded and draws 25% to 50% more current than rated.	Reduce load by reducing speed or gearing in the drive or loading in the drive.
	Voltage above rated.	Motor runs drive above rated speed requiring excessive hp. Reduce voltage to nameplate rating.
	Inadequately ventilated.	Location of motor should be changed.

Figure 13-10: Troubleshooting chart for electric motors. ***(Cont.)***

Symptoms	Probable Cause	Action or Items to Check
Motor overheats or runs hot	Draws excessive current due to shorted coil.	Repair armature coils or install new coil.
	Grounds in armature such as two grounds which constitute a short.	Locate grounds and repair or rewind with new set of coils.
	Armature rubs pole faces due to off-center rotor causing friction and excessive current.	Check brackets or pedestals to center rotor and determine condition of bearing wear for bearing replacement.

Figure 13-10: Troubleshooting chart for electric motors. ***(Cont.)***

Symptoms	Probable Cause	Action or Items to Check
Hot armature	Core hot in one spot indicating shorted punching and high iron loss.	Sometimes full slot metal wedges have been used for balancing. These should be removed and other means of balancing be investigated.
	Brush tension too high.	Limit pressure to 2 to $2\frac{1}{2}$ psi. Check brush density and limit to density recommended by the brush manufacturer.

Figure 13-10: Troubleshooting chart for electric motors. ***(Cont.)***

Symptoms	Probable Cause	Action or Items to Check
Hot armature	Punching uninsulated. Punching has been turned or band grooves machined in the core. Machined slots.	No load running of motor will indicate hot core and drawing high no-load armature current. Replace core and rewind armature. If necessary to add band grooves grind into core. Check temperature on core with thermometer not to exceed 90°C.
Hot commutator	Brushes off neutral.	Reset neutral.

Figure 13-10: Troubleshooting chart for electric motors. ***(Cont.)***

Symptoms	Probable Cause	Action or Items to Check
Hot commutator	Brush grade too abrasive.	Get recommendation from manufacturer.
	Shorted bars.	Investigate commutator mica and undercutting, and repair.
	Hot core and coils that transmit heat to commutator.	Check temperature of commutator with thermometer to see that total temperature does not exceed ambient plus 55°C rise, total not to exceed 105°C.

Figure 13-10: Troubleshooting chart for electric motors. ***(Cont.)***

Symptoms	Probable Cause	Action or Items to Check
Hot commutator	Inadequate ventilation.	Check as for hot motor.
Hot fields	Voltage too high.	Check with meter and thermometer and correct voltage to nameplate value.
	Shorted turns or grounded turns.	Repair or replace with new coil.
	Resistance of each coil not the same.	Check each individual coil for equal resistance within 10% and if one coil is too low replace coil.

Figure 13-10: Troubleshooting chart for electric motors. ***(Cont.)***

Symptoms	Probable Cause	Action or Items to Check
Hot fields	Inadequate ventilation.	Check as for hot motor.
	Coil not large enough to radiate its loss wattage.	New coils should replace all coils if room is available in motor.
Motor vibrates and indicates unbalance	Armature out of balance.	Remove and statically balance or balance in dynamic balancing machine.
	Misalignment.	Realign.

Figure 13-10: Troubleshooting chart for electric motors. ***(Cont.)***

Symptoms	Probable Cause	Action or Items to Check
Motor vibrates and indicates unbalance	Loose or eccentric pulley.	Tighten pulley on shaft or correct eccentric pulley.
	Belt or chain whip.	Adjust belt tension.
	Mismating of gear and pinion.	Recut, realign, or replace parts.
	Unbalance in coupling.	Rebalance coupling.
	Bent shaft.	Replace or straighten shaft.

Figure 13-10: Troubleshooting chart for electric motors. ***(Cont.)***

Symptoms	Probable Cause	Action or Items to Check
Motor vibrates and indicates unbalance	Foundation inadequate.	Stiffen mounting place members.
	Motor loosely mounted.	Tighten holding-down bolts.
	Motor feet uneven.	Add shims under foot pads to mount each foot tight.
Motor sparks at brushes or does not commutate	Neutral setting not true neutral.	Check and set on factory setting or test for true neutral.

Figure 13-10: Troubleshooting chart for electric motors. ***(Cont.)***

Symptoms	Probable Cause	Action or Items to Check
Motor sparks at brushes or does not commutate	Commutator rough.	Grind and roll edge of each bar.
	Commutator eccentric.	Turn, grind and roll commutator.
	Mica high—not undercut.	Undercut mica.
	Commutating pole strength too great causing over-compensation or strength too weak indicating under-compensation.	Check with manufacturer for correct change in air gap or new coils for the commutating coils.

Figure 13-10: Troubleshooting chart for electric motors. ***(Cont.)***

Symptoms	Probable Cause	Action or Items to Check
Motor sparks at brushes or does not commutate	Shorted commutating pole turns.	Repair coils or install new coils.
	Shorted armature coils on commutator bars.	Repair armature by putting into first class condition.
	Open circuited coils.	Same as above.
	Poor soldered connection to commutator bars.	Resolder with proper alloy of tin solder.

Figure 13-10: Troubleshooting chart for electric motors. ***(Cont.)***

Symptoms	Probable Cause	Action or Items to Check
Motor sparks at brushes or does not commutate	High bar or loose bar in commutator at high speeds.	Inspect commutator nut or bolts and retighten and grind commutator face.
	Brush grade wrong type. Brush pressure too light, current density excessive, brushes stuck in holders. Brush shunts loose.	See brushes.
	Brushes chatter due to dirty film on commutator.	Resurface commutator face and check for change in brushes.

Figure 13-10: Troubleshooting chart for electric motors. ***(Cont.)***

Symptoms	Probable Cause	Action or Items to Check
Motor sparks at brushes or does not commutate	Vibration.	Eliminate cause of vibration by checking mounting and balance of rotor.
Brush wear excessive	Brushes too soft.	Blow dust from motor and replace brushes with a changed grade as recommended by manufacturer.
	Cummutator rough.	Grind commutator face.

Figure 13-10: Troubleshooting chart for electric motors. ***(Cont.)***

Symptoms	Probable Cause	Action or Items to Check
Brush wear excessive	Off neutral setting.	Recheck factory neutral or test for true neutral.
	Abrasive dust in ventilating air.	Reface brushes and correct condition by protecting motor.
	Bad commutation.	See corrections for commutation.
	High/ low loose bar.	Retighten commutator motor bolts and resurface commutator.

Figure 13-10: Troubleshooting chart for electric motors. ***(Cont.)***

Symptoms	Probable Cause	Action or Items to Check
Brush wear excessive	Brush tension excessive.	Adjust spring pressure not to exceed 2 to 2½ psi.
	Electrical wear due to loss of film on commutator face.	Resurface brush faces and commutator face. Check for change in brush grade.
	Threading and grooving.	Same as above.
	Oil or grease from atmosphere or bearings.	Correct oil condition and surface brush faces and commutator.

Figure 13-10: Troubleshooting chart for electric motors. ***(Cont.)***

Symptoms	Probable Cause	Action or Items to Check
Brush wear excessive	Weak acid and moisture laden atmosphere.	Protect motor by changing ventilating air, or change to enclosed motor.
Motor noisy	Brush singing.	Check brush angle and commutator coating, resurface commutator.
	Brush chatter.	Resurface commutator and brush face.
	Motor loosely mounted.	Tighten foundation bolts.

Figure 13-10: Troubleshooting chart for electric motors. *(Cont.)*

Symptoms	Probable Cause	Action or Items to Check
Motor noisy	Foundation hollow and acts as sounding board.	Coat underside with sound proofing material.
	Strained frame.	Shim motor feet for equal mounting.
	Armature punching loose.	Replace core on armature.
	Armature rubs pole faces.	Recenter by replacing bearings or relocating brackets or pedestals.
	Magnetic hum.	Refer to manufacturer.

Figure 13-10: Troubleshooting chart for electric motors. ***(Cont.)***

Symptoms	Probable Cause	Action or Items to Check
Motor noisy	Belt slap or pounding.	Check condition of belt and change belt tension.
	Excessive current load.	May not cause overheating but check chart for correction of shorted or grounded coils.
	Mechanical vibration.	Check chart for causes of vibration.
	Noisy bearings.	Check alignment, loading of bearings and lubrication.

Figure 13-10: Troubleshooting chart for electric motors. ***(Cont.)***

Symptoms	Probable Cause	Action or Items to Check
Motor stalls	Wrong application.	Change type or size. Consult manufacturer.
	Overloaded motor.	Reduce load.
	Low motor voltage.	See that nameplate voltage is maintained.
	Open circuit.	Fuses blown, check overload relay, starter and pushbuttons.

Figure 13-10: Troubleshooting chart for electric motors. ***(Cont.)***

Symptoms	Probable Cause	Action or Items to Check
Motor stalls	Incorrect control resistance of wound rotor.	Check control sequence. Replace broken resistors. Repair open circuits.
Motor connected but does not start	One phase open.	See that no phase is open.
	Motor may be overloaded.	Reduce load.
	Rotor defective.	Look for broken bars or rings.
	Poor stator coil connection.	Remove end bells.

Figure 13-10: Troubleshooting chart for electric motors. ***(Cont.)***

Symptoms	Probable Cause	Action or Items to Check
Motor runs and then dies down	Power failure.	Check for loose connections to line, to fuses and to control.
Motor does not come up to speed	Not applied properly.	Consult supplier for proper type.
	Voltage too low at motor terminals because of line drop.	Use higher voltage on transformer terminals or reduce load.

Figure 13-10: Troubleshooting chart for electric motors. ***(Cont.)***

Symptoms	Probable Cause	Action or Items to Check
Motor does not come up to speed	If wound rotor, improper control operation of secondary resistance.	Correct secondary control.
	Starting load too high.	Check load motor is supposed to carry at start.
	Low pull-in torque of synchronous motor.	Change rotor starting resistance or change rotor design.
	Check that all brushes are riding on rings.	Check secondary connections.

Figure 13-10: Troubleshooting chart for electric motors. ***(Cont.)***

Symptoms	Probable Cause	Action or Items to Check
Motor does not come up to speed	Broken rotor bars.	Look for cracks near the rings. A new rotor may be required as repairs are usually temporary.
	Open primary circuit.	Locate fault with testing device and repair.
Motor takes too long to accelerate	Excess loading.	Reduce load.
	Poor circuit.	Check for high resistance.

Figure 13-10: Troubleshooting chart for electric motors. ***(Cont.)***

Symptoms	Probable Cause	Action or Items to Check
Motor takes too long to accelerate	Defective squirrel-cage rotor.	Replace with new rotor.
	Applied voltage too low.	Get power company to increase voltage tap.
Wrong rotation	Wrong sequence of phase.	Reverse connections of motor or at switchboard.
Motor overheats while running under load	Check for overload.	Reduce load.

Figure 13-10: Troubleshooting chart for electric motors. ***(Cont.)***

Symptoms	Probable Cause	Action or Items to Check
Motor overheats while running under load	Wrong blowers or air shields, may be clogged with dirt and prevent proper ventilation of motor.	Good ventilation is manifest when a continuous stream of air leaves the motor. If not, check manufacturer.
	Motor may have one phase open.	Check to make sure that all leads are well connected.
	Grounded coil.	Locate and repair.
	Unbalanced terminal voltage.	Check for faulty leads, connections and transformers.

Figure 13-10: Troubleshooting chart for electric motors. ***(Cont.)***

Symptoms	Probable Cause	Action or Items to Check
Motor overheats while running under load	Shorted stator coil.	Repair and then check wattmeter reading.
	Faulty connection.	Indicate by high resistance.
	High voltage.	Check terminals of motor with voltmeter.
	Low voltage.	Same as for high voltage.
	Rotor rubs stator bore.	If not poor machining, replace worn bearings.

Figure 13-10: Troubleshooting chart for electric motors. ***(Cont.)***

Symptoms	Probable Cause	Action or Items to Check
Motor vibrates after corrections have been made	Motor misaligned.	Realign.
	Weak foundations.	Strengthen base.
	Coupling out of balance.	Balance coupling.
	Driven equipment unbalanced.	Rebalance driven equipment.
	Defective ball bearing.	Replace bearing.
	Bearings not in line.	Line up properly.

Figure 13-10: Troubleshooting chart for electric motors. ***(Cont.)***

Symptoms	Probable Cause	Action or Items to Check
Motor vibrates afer corrections have been made	Balancing weights shifted.	Rebalance rotor.
	Wound rotor coils replaced.	Rebalance rotor.
	Polyphase motor running single phase.	Check for open circuit.
	Excessive end play.	Adjust bearing or add washer.

Figure 13-10: Troubleshooting chart for electric motors. ***(Cont.)***

Symptoms	Probable Cause	Action or Items to Check
Unbalanced line current on polyphase motors during normal operation	Unequal terminal volts.	Check leads and connections.
	Single phase operation.	Check for open contacts.
	Poor rotor contacts in control wound-rotor resistance.	Check control devices.
	Brushes not in proper position in wound rotor.	See that brushes are properly seated and shunts in good condition.

Figure 13-10: Troubleshooting chart for electric motors. ***(Cont.)***

Symptoms	Probable Cause	Action or Items to Check
Scraping noise	Fan rubbing air shield.	Remove interference.
	Fan striking insulation.	Clear fan.
	Loose on bedplate.	Tighten holding bolts.
Magnetic noise	Air gap not uniform.	Check and correct bracket fits or bearing.
	Loose bearings.	Correct or renew.
	Rotor unbalance.	Rebalance.

Figure 13-10: Troubleshooting chart for electric motors. ***(Cont.)***

Index

A

B

C

D

G

H

I

K

L

M

N

O

P

Q

R

S

T